SpringerBriefs in Earth Sciences

For further volumes:
http://www.springer.com/series/8897

John F. Hermance

Historical Variability of Rainfall in the African East Sahel of Sudan

Implications for Development

 Springer

John F. Hermance
Brown University
Rehoboth, MA
USA

ISSN 2191-5369   ISSN 2191-5377   (electronic)
ISBN 978-3-319-00574-4   ISBN 978-3-319-00575-1   (eBook)
DOI 10.1007/978-3-319-00575-1
Springer Cham Heidelberg New York Dordrecht London

Library of Congress Control Number: 2013943572

Printed on acid-free paper

Springer is part of Springer Science+Business Media (www.springer.com)

Preface

A fundamental requisite for the social and economic sustainability of peoples in the semi-arid zone south of the Sahara is that both indigenous cultures and emerging economies need to adapt to the natural constraints imposed by the local environment. In the African Sahel, these constraints are prescribed by available rainfall. It follows that rainfall data, and their analysis, are essential to a variety of concerns in the region, affecting populations ranging from nomadic herders to mechanized farming. Drawing on an analysis of the historical rainfall record, this report describes the basic patterns of distribution and variability of rainfall in the East Sahel.

Many planners are concerned with what the future holds for rainfall in the region due to climate change. The premise of this report is that the best evidence we currently have for the future is based on historical records of the past. Because it is imperative that the general public appreciate the basis and the implications of this assertion, the discussion here—while perhaps containing a few reminders for climate specialists—is directed to a general readership of students, policy-makers, and non-specialist scientists and engineers.

The study focuses on the East Sahel by analyzing more than 100 years of historical rainfall data from three of the few long-term standard World Meteorological Organization (WMO) rain gauge stations in substantially different rainfall settings. From north to south, transecting the Sahel over a distance of several hundred kilometer, the stations with their annual rainfall are Khartoum (130 mm), Kassala (280 mm), and Gedaref (600 mm). These three stations effectively span the rainfall gradient transecting the rolling savanna in the most eastern section of the Sahel, close to the intersection of the border of Sudan with Ethiopia and Eritrea.

The conclusions challenge a popular notion that changing climate—increased drought and desertification—in the East Sahel may have already accelerated the deterioration of its water resources. The analysis here, however, shows that any evidence of a persistent and coherent regional trend of diminishing rainfall is minimal. Quite the contrary, the evidence demonstrates that the fluctuations of climate and weather patterns over the ensuing decades of the past century—at all temporal scales from days to years to decades—profoundly overwhelm any suggestion of a large-scale, coherent decrease (or increase) in rainfall. The implication is that—in terms of naturally induced threats to a community—it is not long-term change, but the highly localized interseasonal, interannual, and multiannual variability of rainfall that poses the greatest and most immediate risk to an agrarian economy struggling to survive in a climate that irregularly vacillates between years of drought and years of flooding.

The analysis is intentionally straightforward and uncomplicated, so that its results may be readily assessed and applied by the widest community. The statistics are quite basic. What the data show for the East Sahel is that over the past century, as much as 96 % or more of the variance in annual rainfall totals is associated with interannual to multidecadal quasi-cyclic fluctuations, and any significant trend is lost in the remaining few percent of the overall variance. At the same time, interannual difference in rainfall by up to a factor of two are not uncommon, and most striking is the asynchronous occurrence of positive and negative extremes in annual totals among stations only a hundred or so kilometers apart. Similar to previous studies in the West Sahel, the data from the East Sahel show that in spite of the significant gradient of annual totals between stations, the magnitude of storms is not very different across the Sahel; the primary difference is in the *number* of storms per season at a site, not necessarily differences in their *intensity*. Most of the storms at a respective station are of low intensity, but most of the total rainfall in a year is contributed by the fewer higher intensity events. The discussion explores in some detail common assumptions in the literature regarding the month(s) of maximum rainfall, and the month(s) that contribute most greatly to the overall variability of annual rainfall. Systematic differences in the length of the monsoon season across the Sahel range from 3 months in the north to 4 months in the south. Whereas the relative interannual variability of rainfall *increases* as one moves from the higher rainfalls in the south to the lower rainfalls in the north, the actual, or absolute interannual variability *decreases*. Analytical models are developed to represent the expected seasonal distributions of rainfall that are relevant to both agricultural planning and runoff forecasting. Daily data are used to explore expected short-term rainfall extremes, not only the expected magnitudes of the larger storm events, but their frequency, along with the frequency of the equally important storm-free intervals, which, of course, provides information on the expected duration of interseasonal and interannual dry periods or droughts.

The thematic thread woven throughout this discussion is the singular influence that the spatial and temporal variability of rainfall has on local communities. The evidence suggests that local communities might experience catastrophic flash floods or midseason droughts that go largely undetected by the international aid community, perhaps by even the national government. While global climate models have successfully reconstructed some of the broad regional features of climate in the Sahel, the detailed patterns evident in the historical record suggest that the local behaviors of climate appear to be much more challenging. Yet these local events, because of their extreme nature, are of enormous significance to a local population.

It is the view of the author that the international community is ill-prepared to either identify such extreme events at the scale on which they most commonly occur, or, if detected, the international community is ill-posed to implement emergency procedures to mitigate their impact. In fact, considering the sparsity and neglect of long-term, surface-based rain gauges in the region, there are few areas of the Sahel where an adequate database exists so that sufficient planning can even begin. This report provides examples of some of the metrics that are not only useful for identifying rainfall extremes, but are essential for engineering design and routine water management. In the context of this report, a substantial reserve of local knowledge lies in the local rainfall records that have not yet found their way into the international archives. It is critical that the international scientific community—if it is to merge its expertise with the on-the-ground needs of local experts—have better access to the data that define the nature of the problems at hand. Climate modelers are at a crippling disadvantage if they do not have the data to model.

The subject matter in this text has evolved from courses in qualitative and quantitative physical hydrology developed by the author over his years at Brown University, a relatively small, liberal arts university–college within which strong programs in the physical, biological, and engineering sciences thrive. Special thanks to those students—humanists and scientists, undergraduate and graduate, poets and engineers—who connected, and who continue to connect, each in their own way, the basic elements of hydrology with the lives of local peoples.

Rehoboth, MA, USA, May 2013   Jack (John F.) Hermance

Contents

Chapter 1
Introduction

Abstract Water—specifically rainfall—is the ultimate determinant of sustainability for indigenous and developing communities in the semi-arid African Sahel. Rainfall in this part of the world is associated with the annual south-to-north migration of the Intertropical Convergence Zone (ITCZ) and its attendant rain belt. International agencies are coming to realize that African lifestyles cannot be managed from afar, or by persons not familiar with the indigenous cultures. Local knowledge is increasingly appreciated by the planning community, and this report insists that local knowledge must include knowledge of local climate, particularly for the Sahel where rainfall variability in space and time plays such a capricious role in everyday life. This and the following chapters use 100 years of data from each of three standard WMO rain gauge stations in the East Sahel. From north to south, they are Khartoum (KHA: annual prcp = 130 mm), Kassala (KAS: annual prcp = 280 mm), and Gedaref (GED: annual prcp = 600 mm), so that the stations lie along a relatively intense northwest to southeast rainfall gradient of approximately 2.2 mm/km/yr. The analysis underscores the type of extreme seasonal and interannual spatial and temporal rainfall variability that has historically posed, and will continue to pose, significant and immediate threats to sustainable communities in the Sahel. Whereas some policy analysts are concerned with hypothetical scenarios predicted 50 years or more into the future, the evidence in this report shows that the clear and present danger to the social fabric of indigenous cultures is from the ongoing occurrence of the types of climate extremes regularly experienced in the historical records of the past.

1.1 Water Availability in Sub-Saharan Africa

The availability of water in Africa south of the Sahara—whether too little, resulting in droughts, or too much, resulting in floods—is a critical determinant in the well-being of local communities and the health of natural ecosystems.

J. F. Hermance, *Historical Variability of Rainfall in the African*
East Sahel of Sudan, SpringerBriefs in Earth Sciences,
DOI: 10.1007/978-3-319-00575-1_1, © The Author(s) 2014

By "available water", in this part of the world, one means, of course, rainfall; and by rainfall one is invariably referring to the rains seasonally delivered to the region by the African monsoon (Nicholson 2011; Spinage 2012) and all that this appellation implies about the geographic distribution and variability of precipitation in this area (Fig. 1.1).

Much research over the decades has focused on rainfall and water availability for the African continent, with a comprehensive, non-specialist overview of the science, and policy implications, provided by the Africa Water Atlas (UNEP 2010). The discussion that follows here builds on that research by taking a very basic view of the interannual and interseasonal behavior of the monsoon in the East Sahel. It draws on historical, public-domain rainfall records archived by the international community through the World Meteorological Organization (WMO) and the Global Historical Climate Network (GHCN) to bring into focus certain aspects of the hydroclimate that seem to have escaped the close attention they deserve, but nevertheless have considerable, if not overwhelming, relevance to a larger community of stakeholders. These include not only climate scientists, but environmental managers, potential developers, international aid agencies, as well as local small-scale farmers and transhumant pastoralists. The evidence in this report will demonstrate that the primary threat to sustainable water supplies in the East Sahel—and by analogy, the Sahel in general—is not whether a community has a mean annual rainfall of 200 or 600 mm, or even if, as suggested by some, the Sahel is experiencing a long-term trend of diminishing rainfall, which, in fact, is challenged by this analysis. Rather, the principal threat to community sustainability, at least from natural factors of

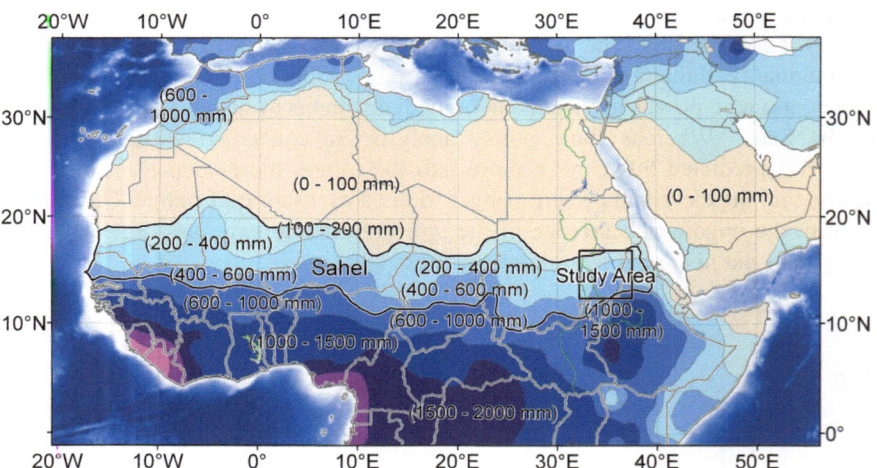

Fig. 1.1 Annual precipitation totals (mm) for the climate normal period: 1951–1980 (Deichmann and Eklundh 1991). The African Sahel is that zone south of the Sahara having annual precipitation in the range of 100–600 mm as outlined by heavy contours in the figure. The study area in this report is shown by the rectangular area in the East Sahel spanning the border areas of Sudan, Eritrea, and Ethiopia

climate, is the inexorable imprint of interseasonal, interannual, even multi-decadal, variability of local rainfall.

The general latitude distribution of annual precipitation over Africa north of the equator (Fig. 1.1) is commonly acknowledged to be driven by seasonal monsoons associated with the annual latitudinal shift of the inter-tropical convergence zone (ITCZ) as it, and its associated equatorial Hadley cell systems, move northward and southward in response to the seasonal solar cycle (Nicholson 2011; Spinage 2012; Holden 2005). The mid-latitude high pressure systems associated with the moisture-depleted descending limbs of the Hadley cells are largely credited— particularly in the winter months—for the arid to semi-arid conditions in the Sahara to the north and South Africa to the south. In the summer months, along the southern edge of the Sahara, the dry, overland northeasterly trade winds from the Middle East compete with the humid winds from the southwest, southeast, and south along the Intertropical Front (ITF), often used by international monitoring agencies as an operational demarcation of the current northernmost extent of substantive monsoon rainfall (CPC 2012).

The Sahel is the transition zone south of the Sahara (Fig. 1.1, 1.2) associated with the north-south gradient in precipitation from the arid Sahara to the north, to the humid equatorial zone of central Africa to the south, having annual monsoonal precipitation ranging from 100 mm along its northern edge to 600 mm along its southern edge. The impact of this precipitation pattern on surface vegetation during the boreal (northern) summer is illustrated by the satellite image in Fig. 1.2 of the normalized differential vegetation index (NDVI) for September, 1992. NDVI is basically a "greenness" index using satellite observations in the optical and near-infrared spectral bands to assess the status of surface vegetation, usually at

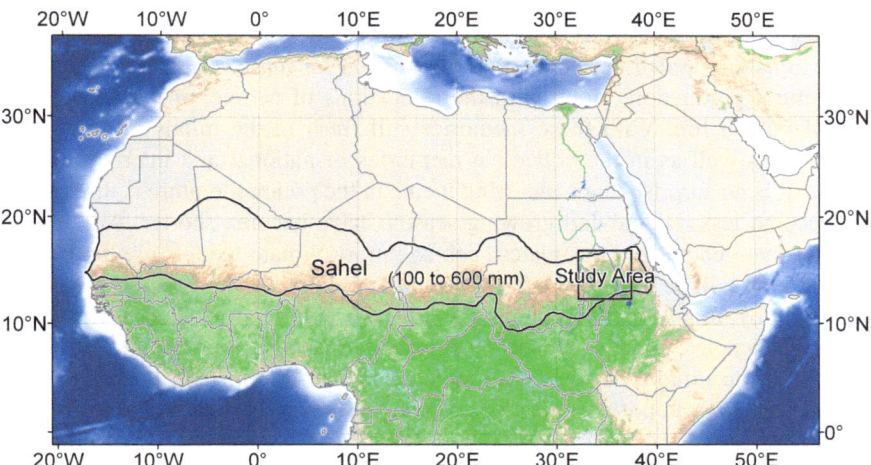

Fig. 1.2 Vegetation map from USGS NDVI database for September 1992. The Sahel boundaries are from Fig. 1.1. The location of the study area for this report is shown by the box in the East Sahel

dekadal (10 day) to monthly intervals. NDVI maps of the Sahel have proven to be essential tools for monitoring the condition of landcover—and a proxy for rainfall—in the Sahel (Tucker et al. 1986; Funk and Brown 2005; Verdin et al. 2005).

1.2 Refocusing Climate and Water Policy from Regional to Local Scales

A considerable body of literature has developed concerning the proper management of the water resources of the Sahel, with many of the plans fostered by the United Nations Environment Programme (UNEP) and non-government organizations (NGOs)—see for example UNEP (2003), Kandji et al. (2006), the Africa Atlas (UNEP 2008), Ngigi (2009), and the Africa Water Atlas (UNEP 2010). The international community is coming to realize, often through prodding by local stakeholders, that African lifestyles cannot be managed from afar, or by persons not familiar with the indigenous cultures and local knowledge. In the past, ambitious development plans, particularly in the area of agriculture, catastrophically failed due to lack of community participation (El Gamri 2004). Sustainable usage of local land and water is an evolving process that works best if local peoples with indigenous knowledge are directly engaged in planning and implementation (Sulieman et al. 2012). Since water—namely rainfall—plays such a determining factor in the well-being of local communities, it seems inevitable that planning should adjust its perspective to account for the natural patterns of rainfall at the local level.

By delegating the planning and management of water policy to local communities, one might anticipate that the respective responses would be influenced by memories of the historic Sahel droughts of the 1970s and 1980s—what the UNEP refers to as the "Sahel desiccation"—responsible for massive losses of livestock, agricultural production, along with the relocations of people, and ultimately the loss of human life. With these memories still fresh in the minds of local communities, as well as in the collective memories of national and international aid groups, it is no surprise that a recurring theme in the policy literature is that rainfall has become less reliable and growing seasons have become shorter. And it is no surprise that generalized postulates such as "most climate models predict that the Sahel region will be drier in the 21st" (Kandji et al. 2006), are becoming accepted scenarios by many in the potentially affected communities and international planning agencies. What some see as an extended drought through the last half of the past century might be easily extrapolated as ever-diminishing rainfall into the next century. An important question, of course, is whether such claims are well founded in fact, or at least are they well founded on the scale at which they are implied.

The UNEP Africa Atlas of Our Changing Environment (UNEP 2008) notes that more than 300 million people in Sub-Saharan Africa currently face water scarcity, and expects that the numbers of those experiencing water shortages will increase substantially by 2050. A number of reports warn against impending major food shortages and increased desertification due to climate change, and draw on information from the scientific community to address these issues on broad regional and national scales (Toulmin 2009). What role can science play? The problem is that the scientific community is still developing the tools (e.g. computer models) and observational databases (e.g. remote sensing satellites) to understand the underlying processes on subcontinental scales at resolutions of many tens to hundreds of km or more (GOSIC 2012), not on the scale of 10 km or less, which is critically needed for planning by local communities. This report will show that patterns of climate—rainfall—at the local community level suggests a substantially different story from that implied by aggregated climate patterns reconstructed on the subcontinental scale.

While a broad-brush approach is essential background for understanding climate—particularly for understanding the fundamental driving mechanisms for global-scale climate variability—at some point these problems become local, so that to discuss strategies to locally mitigate the impacts of climate change, one needs to ask two questions: "Change from what?"; and "What is the character of change in the past?" This is where the assessment of historic precipitation data becomes essential.

Practically speaking, water availability is defined and constrained by three factors: (1) Amount of rainfall; (2) Variability of rainfall amounts; (3) The engineered development and sustainability of potable water delivered to local communities. In this report, we propose to step back from the engineering and management aspects to first get a better idea of what, exactly, there is to manage by way of available water (rainfall) and to ask what are the natural characteristics of the resource itself. Where does it rain? When does it rain? And how much does it rain? While such questions have received considerable attention on the subcontinental and national scales (for example, Kraus 1977; Nicholson 1986; Mahé and Paturel 2009; the Africa Water Atlas, UNEP 2010), in this report, we address them on a more site-specific scale for a local study area in the East Sahel. Discussion of the third item—appropriate engineering design and agricultural management—is deferred since substantial advances on *sustainable* development cannot be made without due awareness of the first two items: the fundamental constraints that the natural environment places on the availability of water, in particular on the time and spatial scale of local concern to farmers, managers, and developers.

This report will show that, for proper planning, the characterization of the regional hydroclimate involves significantly more than determining the mean annual rainfall. The latter—the mean annual rainfall—is a metric referred to by climatologists as the "climate normal" for the respective region; and is referred to by statisticians as the "expected" or the "expectation" value for annual precipitation. Historically, years having precipitation totals above the normal or expected value are "wet years," often flood years, and years having precipitation totals less than the normal or expected value are "dry years," often drought years. But as so eruditely observed by the late Eric Kraus:

There are no absolute criteria for drought; the dryness of the desert does not wilt the thorn bush or the cactus, and is not experienced as a disaster by Bushman or Bedouins (Kraus 1977).

He goes on to say that, "by definition droughts are anomalies—deviations from a rainfall regime to which people, plant and animals have adapted as the local norm." The same might be said of wet years, as flash floods and major deluges locally disrupt the normal lifestyles of the transhumant nomadic herder or wash out the cultivated crops of the village pastoralist. Local communities have adapted to the normal climate, from semi-arid to semi-humid, but it is the anomalies that destroy the patterns of normal life. Hulme (2001) asserts that

There is no such thing as 'normal' rainfall in the Sahel. What matters fundamentally is not whether the mean annual rainfall is 200, 400, or 600 mm, but the spectrum of rainfall variability in time and space.

He argues that, whereas this variability is well known to indigenous African societies, it was largely ignored in the colonial and post-colonial eras until the last half of the Twentieth century, when in the 1970s, the Sahel experienced its great multi-decadal drought. He contends that the variability of rainfall relative to some arbitrary background, steady-state value, is often more important than the steady-state value—the climate "normal"—itself, a sentiment that is increasing echoed in the recent ecological literature (Spinage 2012). It is this interannual and inter-seasonal variability of rainfall for a local area of the Sahel that will be addressed in the present report.

1.3 Scope of the Discussion

The discussion is divided into six chapters, throughout which one will find the basic analysis and the salient conclusions relative to the topic at hand. The first chapter—the Introduction—describes the objectives of this investigation, emphasizing the importance of merging information from climate science with sustainable policies of water management at the local level, including a general background of the study area relative to local precipitation patterns in space and time. The second chapter will employ century-long time series of annual rainfall, from the beginning of the 1900s to 2009, along with the statistics of annual values over this time period, to describe the departure of these annual totals from their long-term expected values. The interannual variability of annual totals will be compared with those from other regions of the Sahel to better understand the scale of spatial variability involved. The third chapter will rely largely on monthly data to explore interannual and interseasonal variability of rainfall, concluding with a discussion on the oft-stated notions that August is typically the month of maximum rainfall, and is the month whose variability in rainfall contributes most greatly to interannual patterns of variability. Next, in the fourth chapter, daily precipitation data from the East Sahel—and as expected, according to GCOS (2004), for only limited periods—will be used to characterize rainfall frequency and intensity. Last in this chapter is the development of seasonal models for several sites in the East Sahel to represent the temporal distribution of rainfall throughout the year. Such models have proven to be useful elsewhere for runoff studies, and seasonal streamflow generation, although in this paper they are used primarily to show the interstation variability in the timing and extent of the monsoon season. The fifth chapter will move on to explore certain aspects of the duration and frequency of discrete multiday storm events. Along with the nature of the storm events them-selves, are questions on the nature of interstorm dry periods. Thus, basic statistics are developed on the temporal (and spatial) variability of the intraseasonal dry periods—that is to say, the dry periods during the monsoon or growing season. Also discussed is the inevitable interannual hiatus in rainfall—or the interannual dry period—typical for this region of the African monsoon. The sixth and final chapter summarizes selected findings of this investigation in the context of the essential role that archived data and their analyses play in developing a better understanding of the relationship between local hydroclimate variability in the Sahel and the immediate needs of local communities.

1.4 Long-Term Precipitation Patterns South of the Sahara

The first-order, fundamental physics of the intra-annual, meridional march of the ITCZ is relatively well understood from ground-based data, satellite observations, atmospheric physics, and numerical modeling (Nicholson 2009, 2011; Biasutti et al. 2008). A recent numerical modeling experiment by Suzuki (2011) assessing the agreement among the results of 22 different coupled general circulation models, is one of the latest to reaffirm that the seasonal procession of the ITCZ over Africa is largely driven by solar insolation. The seasonal migration of the sub-solar point ranges from 23° S latitude in June to 23° N latitude in December, drawing along the lagging tropical rainbelt so that in early August its northern edge is at approximately 15° N latitude, and in late January and early February, its southern edge is an irregular perimeter at roughly 15° S latitude. However, details on the interannual and intraseasonal spatial and temporal variability of rainfall based on observational data are another matter (Bell and Lamb 2006; Nicholson 2009, 2011; Jurkovica and Pasaric 2012). For some years, climate workers in and outside of Africa have emphasized that, unlike the marine areas, the ITCZ over Africa does not concatenate with its associated tropical rain belt (Thompson 1965; El Tom 1975; Nicholson and Grist 2001; Biasutti, et al. 2008; El Gamri et al. 2009). The convergence of surface winds associated with the ITCZ typically occurs several hundred kilometers to the north of the associated rain belt. Nicholson (2009, 2011) has shown that the atmospheric couplings between the rainbelt and the ITCZ for the humidity-starved ecosystems south of the Sahara are somewhat more complicated than classical models—such as those for the humidity-rich open oceans—would suggest. In addition, land surface conditions complicate the matter. It is now understood that the long-term multi-decadal continental scale precipitation patterns, driven by a composite of global climate teleconnections in the atmosphere forced by sea surface temperatures (see, for example, Giannini et al. 2008), are strongly overprinted by the effects of topography and landcover, among other regional and sometimes quite local factors (El Tom 1975; Riddle and Cook 2008; El Gamri et al. 2009).

Much of the research over the past decades has tended to generalize patterns of observational surface-based data to better understand climate processes—particularly precipitation—on a continental to subcontinental scale. Conveniently, this also serves the secondary purpose of aggregating or up-scaling surface data to accommodate the relatively coarse-scales required by contemporary global circulation models (GCMs). At the scale and time frame to which they are applied, the aggregated observations and GCM predictions agree reasonably well, at least for the principal behaviors of past scenarios. Arguably, however, the aggregated observations might be minimizing aspects of local variability in climate (Agnew 2000; Ali and Lebel 2008), particularly in the amounts and timing of rainfall that are quite significant at the local scale. Perhaps the principal point to take away from the analysis in this report is to remind the climate community, as well as water planners, potential developers and farmers, of the magnitude and scale of historical interannual and interseasonal rainfall variability in a region of the Sahel that has been relatively understudied, but is playing an increasing role in world affairs.

1.5 Hydroclimate Setting of the Study Area

This report will focus on historical rainfall in the East Sahel, primarily the area of Sudan shown in Fig. 1.3 close to the intersection of the national boundaries of Sudan, Eritrea, and Ethiopia. Its objective is that, working on the spatial scale of a few, select precipitation gauges (referred to by meteorologists as "hyetographs"), but drawing on the temporal scale of long-term, high-quality time series, its analysis might sharpen the resolution of certain patterns identified in the prior work of others, both observationally (Hulme 1990, 2001; Nicholson 1989; Nicholson and Selato 2000; Elagib and Elhag 2011), as well as from the point of view of mesoscale atmospheric numerical modeling (Riddle and Cook 2008).

The color-coded zones in Fig. 1.3 correspond to mean annual precipitation provided by the UNEP-GRID GIS database for the climate normal period 1951–1980 (Deichmann and Eklundh 1991). The precipitation range for each zone is given in mm per year. The location and names of selected WMO precipitation stations are also shown in Fig. 1.3, with a summary of the principal data on each station provided in Table 1.1. There is clearly a strong north to south climatic gradient in annual precipitation totals from the Sahara, north of Khartoum at the edge of the Sahara, south to Gondar in the Ethiopian Highlands. The pattern is similar to that for the West Sahel, where Balme et al. (2006) report a north-to-south gradient in annual precipitation totals of approximately 1 mm/km for the area of Niamey, Niger.

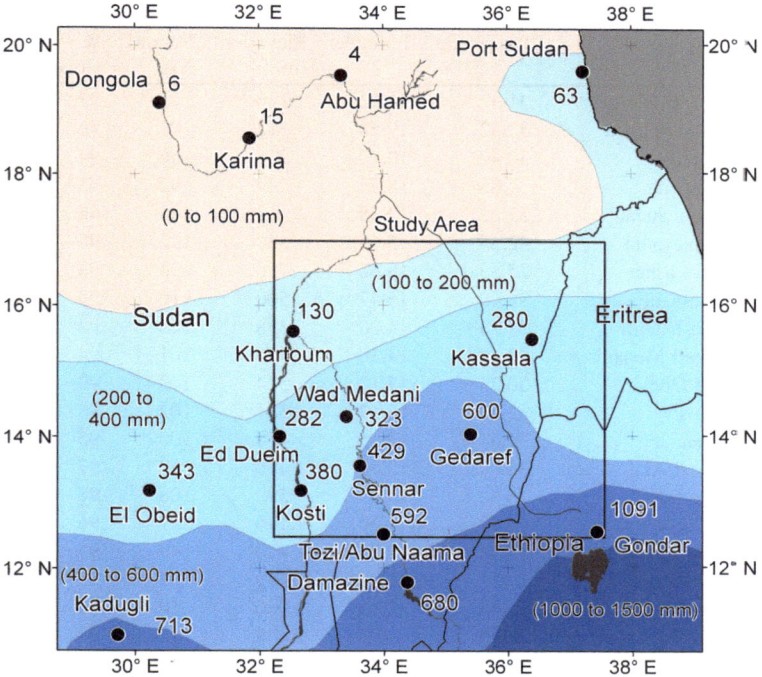

Fig. 1.3 Location of the study area in East Sudan (rectangular area: 1080 km NS by 1080 km EW) and long-term (>100 yr) precipitation stations having current (thru 2009) records. The number next to each station is the respective long-term median annual total precipitation in mm from Table 1.1. The colored contours in the background are the nominally 30-year mean annual precipitation contours for 1951–1980 produced by Deichmann and Eklundh (1991) through the UNEP-GRID program. (Respective ranges for the contours are shown in parentheses)

1.6 Interseasonal Variability of Rainfall

In addition to the strong spatial gradient of annual precipitation shown in Fig. 1.3, there is the strong intra-annual, seasonal variation of precipitation shown in Fig. 1.4 due to the interseasonal meridional shift of the ITCZ about the equator, and its local impact on the monsoon cycle. Patterns of rainfall, particularly south of Khartoum, are significantly influenced by weather patterns deflected around and over the Ethiopian Highlands to the southeast (El Tom 1975; El Gamri et al. 2009). The modeling study of Riddle and Cook (2008) confirms that, along with the transcontinental southwesterly winds carrying moisture from the South Atlantic, the moisture-laden weather systems entering from the Indian Ocean through the Turkana depression south of the Ethiopian Highlands, and circulating around the southwest corner of the latter, are significant players in the development of convective systems over the plains of East Sudan. (Local elevation contours are shown

Table 1.1 Median of total annual precipitation (mm) through 2009 for selected stations

STA_ID	Station	Longitude	Latitude	Elevation (m)	Record Yrs	% Coverage	Prcp (mm)
62600	Wadi Halfa	31.32	21.92	126	73	92	0
62640	Abu Hamed	33.32	19.53	312	102	96	4
62650	Dongola	30.40	19.10	226	66	94	6
62660	Karima	31.85	18.55	249	92	93	15
62641	Port Sudan	37.22	19.58	2	105	96	63
62721	Khartoum	32.55	15.60	380	112	98	130
62760	El Fasher	25.33	13.62	730	94	98	237
62730	Kassala	36.40	15.47	500	110	99	280
62750	Ed Dueim	32.33	14.00	378	106	92	282
62751	Wad Medani	33.40	14.30	405	101	94	323
62771	El Obeid	30.23	13.17	574	109	99	343
62772	Kosti	32.67	13.17	381	102	98	380
62762	Sennar	33.62	13.55	418	100	95	429
62795	Tozi/Abu Naama	34.00	12.50	450	59	97	592
62752	Gedaref	35.40	14.03	599	108	98	600
63471	Dire Dawa	41.87	9.60	1,146	59	96	609
62805	Damazine	34.38	11.78	470	49	95	680
62810	Kadugli	29.72	11.00	499	98	95	713
63331	Gondar	37.43	12.53	1,966	70	70	1,091
63450	Addis Abba	38.75	9.03	2,408	112	98	1,042
63402	Jimma	36.83	7.67	1,676	57	93	1,438

in Fig. 1.5.) Thus, while much can be learned and generalized from the extensive monsoon studies in West Africa (such as those by D'Amato and Lebel 1998; Bell and Lamb 2006; Fink et al. 2006; Nicholson 2009), insight into the nature of precipitation patterns for the East Sahel call for a locally focused analysis.

For example, the time series of monthly values for Kassala shown in Fig. 1.4 shows significant interannual to interdecadal fluctuations, punctuated by sharp monthly spikes and intraseasonal dry periods. The question of the spatial coherency of these patterns on the local scale is dealt with in Chap. 3.

Particularly noteworthy in Fig. 1.4 are the major storm and consequent flood events of 1964, 1974, and 1988, and the regional drought events of 1966, 1984, and 1985. Such intraseasonal variability of rainfall is a major concern to pastoral farmers, emergency responders, and local planners. An unheralded local flash flood from a single storm, or a 20-day local drought, might well be as devastating to a local community as is the regional or national event that captures the attention of the international media. This report will employ straightforward means to describe relatively local, short-term, intraseasonal rainfall variability in the context of longer term, regional hydroclimate patterns.

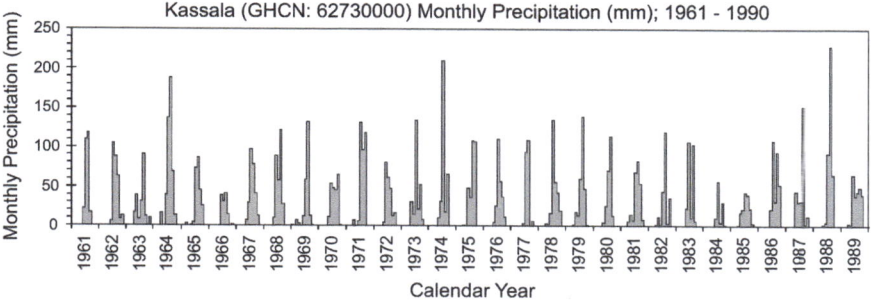

Fig. 1.4 Monthly precipitation data illustrating the annual monsoon cycle for a typical station in the eastern Sahel. The Kassala (GHCN 62730000) station is a member of the GOSIC (NOAA) GSN network.[1] These data—showing a clear single annual monsoon cycle mode—are selected from the current 30 year WMO Climate Normal Period: 1961–1990

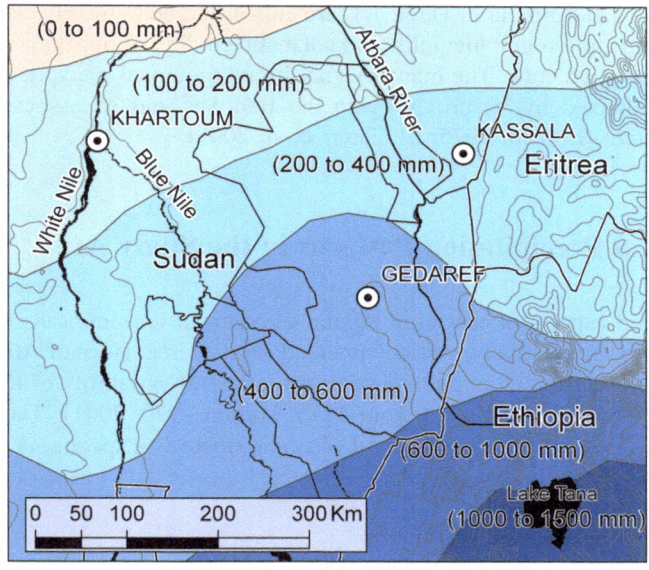

Fig. 1.5 Relative locations of the three precipitation stations used for detailed analysis. From north-to-south parallel to the gradient in precipitation: Khartoum (KHA), Kassala (KAS) and Gedaref (GED). Light-grey contours are for elevation at 200 m intervals. Precipitation zones from UN-EPA. (Absolute elevations of the stations are provided in Table 1.1.)

---

[1] The GSN (Global Systems Network) is a global network of approximately 1,000 stations selected by the Global Climate Observing Systems (GCOS) program from the much larger network of WMO meteorological stations. Precipitation data from the GSN is intended to provide high quality observations from surface-based land stations with a spacing of 2.5–5 degrees of latitude for a number of international missions. (http://gosic.org. Accessed 15 May 2012.)

1.7 Precipitation Patterns for the Sahel of East Sudan: Annual Expected Totals

A subset of the entire suite of rain gauge stations in the Sudan/Ethiopia/Eritrea transboundary area used in this study and listed in Table 1.1, are shown in Fig. 1.3. As mentioned above, the single numeric value next to each station is the median annual precipitation value from Table 1.1 for the entire period of record from the early 1900s (or before) through 2009 superposed on the contoured database for the mean annual precipitation for the climate normal period 1951–1980 from the UNEP/GRID project. The figure shows that, qualitatively, the 30-year, 1951–1980 gridded database of the UNEP accords reasonably well with the 100-year, long-term median annual precipitation at each station.

Note that the *median* value is preferred here as a metric for the *expected* annual value, since the *mean* value for a station tends to be biased by a few outlier events (Tukey 1977). Note also that the annual precipitation at Gondar in Fig. 1.4, as well as Addis Abba and Jimma in Table 1.1, are enhanced due to the effects of induced precipitation from orographic uplift and adiabatic cooling of moist air masses over the Ethiopian Highlands. The intensity and timing of annual precipitation for Port Sudan are affected by its proximity to the Red Sea and its associated marine weather pattern (El Tom 1975; El Gamri et al. 2009).

1.8 Annual Precipitation Patterns in the Study Area

While general aspects of the more regular interannual and intraseasonal precipitation patterns are relatively well documented for the Great Horn of Africa (GHA), most workers recognize the gross spatial and temporal variability of the monsoon rains throughout this region (El Tom 1975; Hulme 1990, 2001). Therefore, one might argue, that for purposes of local planning, there is a specific need for local information.

1.8.1 Location of Study Area

Our focus area is the State of Gedaref, Sudan, because of its location in the easternmost section of the African Sahel (Fig. 1.5), as well as its role as a major source of rainfed agricultural products for the region. To illustrate the seasonal behavior, as well as to analyze the longer term annual patterns of precipitation for the area, we will use data from the three precipitation gauges shown in Fig. 1.5, with station details provided in Table 1.1.

The reference stations are Khartoum (KHA: GHCN 62721), Kassala (KAS: GHCN 62730), and Gedaref (GED: GHCN 62752). Both Khartoum and Kassala are members of the GOSIC (NOAA) GSN network (GOSIC 2012), but Gedaref is not. As in the case for Fig. 1.3, the precipitation range in Fig. 1.5 for each color-coded zone is given in mm per year. Each of the three stations has a period of record longer than 100 years, and at least 98 % coverage. Elevation contours at 200 m intervals are shown by the light-grey lines, and while the latter are unla-belled to minimize clutter in the figure, the contours may be referenced to the absolute elevation of each precipitation station given in Table 1.1.

1.8.2 Regional Gradient of Annual Rainfall

Relative to the respective annual totals at the three stations—Khartoum (KHA; $P_{annual}^{(median)} = 130$ mm), Kassala (KAS; $P_{annual}^{(median)} = 280$ mm) and Gedaref (GED; $P_{annual}^{(median)} = 600$ mm)—Fig. 1.5 shows the pronounced northwest-to-southeast gradient in precipitation. A simple estimate of the spatial dependence of rainfall might be to compare annual rainfall at Khartoum (130 mm) to that at Gedaref (600 mm), where the oblique distance from GED to KHA is approximately 345 km, with a difference in annual precipitation of (600–130) = 470 mm. This would lead to a directed difference of 1.4 mm/km in a direction to the northwest. However, using data from all three stations to estimate a spatial gradient should lead to a more representative value. By definition, the *gradient* is the spatial derivative of a variate that is projected in the direction along which the variate has the greatest rate of change. For the three stations, this is determined to be 2.2 mm/km along a profile directed 8° west of north in Fig. 1.5. Projecting the location of the three stations onto a common profile in this direction with Gedaref (GED) at the origin (0 km), then Kassala (KAS) is at 147 km to the northwest (which might be rounded to 150 km) and Khartoum (KHA) is at 216 km (which might be rounded to 220 km). These locations can be thought of as effective distances of the respective stations along a profile transecting the expected distribution of rainfall during the monsoon season for the region. The stations are thus approximately 1° apart.

1.8.3 General Character of Seasonal Precipitation

The time series in the above Fig. 1.4 illustrates the strong interannual and inter-seasonal variability of precipitation for a single member station of our study: Kassala (KHA). Fig. 1.6 compares seasonal precipitation patterns for Kassala with those of the other two stations: Gedaref (GED) and Khartoum (KHA). The figure shows that precipitation for our three reference stations is usually limited to the months of May (M), June (J), July (J), August (A), September (S), and October (O).

In fact, if it is not already apparent in Fig. 1.6, later chapters will show that the principal months for GED and KAS are JJAS, and for KHA the principal months are JAS. This accords with annual patterns in the West Sahel (Nicholson 2009).

Fig. 1.6 Monthly median
precipitation from all years
during the 100+ year-long
period of record. Note that
the number of the month is
centered on the calendar
period

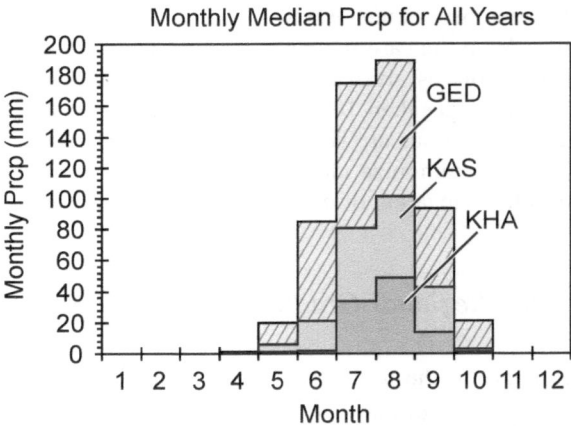

As evident in Fig. 1.6, the three stations in the East Sahel typically have their
expected maximum monthly total during August, as is also the case for the continental West Sahel, although this view is refined in Chap. 3.

1.8.4 Interannual Variability

What is critical to keep in mind throughout this report is the interannual variability
of rainfall, an example of which is provided in Fig. 1.7, which illustrates the
variability of monthly rainfall for one of the study sites: Kassala, a mid-sized
community in the East Sahel, having a mixed urban and pastoral culture. Monthly
data have been selected from the WMO Climate Normal Period: 1961–1990.

For the 30 years of data in Fig. 1.7, there will be 360 (30 yr × 12 mo/yr)
monthly values. In the figure, these 360 values are separated (binned) by month, and
the monthly mean is computed, along with other statistics. The monthly mean for
each month—determined using 30 samples from the 30-year period of record—is
shown by the shaded histogram in Fig. 1.7. Next, the samples within each monthly
bin are internally ranked within that bin by increasing value, and the percentage of
samples (1 thru 30) falling below a monthly value are computed. This percentage is
referred to as a "*quantile*" by statisticians, and is the probability that a monthly
value will be less than the assigned value. Fig. 1.7 shows the median of the monthly
values, which is, of course, the 50 % quantile denoted by Q50; formally, 50 % of the
30 samples for the respective month either equal or fall below this value, and 50 %
of the 30 samples for the month fall above this value. The shaded histogram in the
figure should resemble the analogous histogram for Kassala (KAS) in Fig. 1.6, but
not exactly, since the period of record is different, and the median is used as the
statistical metric in Fig. 1.6, rather than the mean as in Fig. 1.7.

The important points to take away from Fig. 1.7 are the intrinsic variabilities of
rainfall on two time scales: (1) intra-annual variability, namely monthly differences

Fig. 1.7 Interannual
variability of monthly data
for Kassala (KHA), Sudan.
(WMO Stn # 62730.)

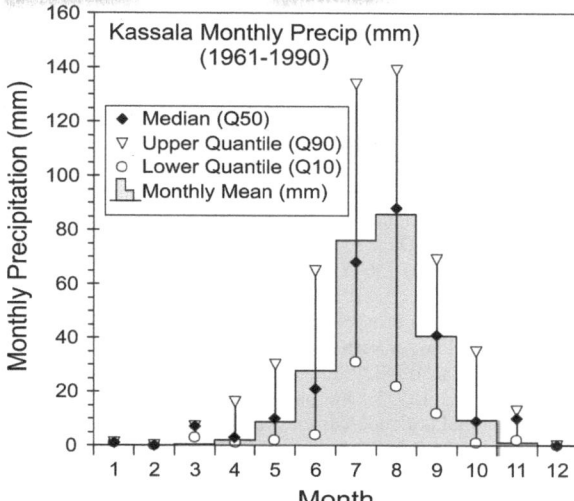

in rainfall—expected and unexpected—throughout the year; and (2) interannual variability as evidenced by the vertical brackets on each month. In terms of implications of interannual variability of monthly rainfall, for example, Fig. 1.7 shows total monthly rainfall in August is less than 20 mm for 10 % of the years; and greater than 140 mm for 10 % of the years. If we define these 10 % limits as *"extreme values,"* then the implications for water management is that, for Kassala, August will experience extreme rainfall 20 % of the time: one out of 5 years, the countryside will have too little, or too much rainfall.

The principal assertion of this report is that these interannual (and interseasonal) fluctuations in rainfall have historically posed, and will continue to pose significant and immediate threats to sustainable communities in the Sahel. Whereas some members of the community are concerned with hypothetical scenarios predicted 50 years into the future from long-range forecasts of global change—namely, warming—the evidence shows that the clear and present danger is from the type of persistent short- to long-term variability already palpable in the climate records of the past.

References

Agnew CT (2000) Using the SPI to identify drought. Drought network news (1994–2001). Paper 1 : http://digitalcommons.unl.edu/droughtnetnews/1. Accessed 29 Apr 2012

Ali A, Lebel T (2008) The Sahelian standardized rainfall index revisited. Int J Climatol Published online in Wiley InterScience, (www.interscience.wiley.com) doi: 10.1002/joc.1832

Balme M, Vischel T, Lebel T, Peugeot C, Galle S (2006) Sahelian water balance: Impact of the mesoscale rainfall variability on runoff. Part 1: rainfall variability analysis. J Hydrol 33:336–348

Bell MA, Lamb PJ (2006) Integration of weather system variability to multidecadal regional climate change: the West African Sudan-Sahel zone, 1951–98. J Clim 19(20):5343–5365

Biasutti M, Held IM, Sobel AH, Giannini A (2008) SST forcings and Sahel rainfall variability in s mulations of the twentieth and twenty-first centuries. J Clim 21:3471–3486

CPC (2012) Climate prediction center, African desk. http://www.cpc.ncep.noaa.gov/productss/african_desk/cpc_intl/. Accessed 28 Sept 2012

D'Amato N, Lebel T (1998) On the characteristics of the rainfall events in the Sahel with a view to the analysis of climatic variability. Int J Climatol 18:955–974

Deichmann U, Eklundh L (1991) Global digital data sets for land degradation studies: a GIS approach. GRID case study series no. 4; UNEP/GEMS and GRID; Nairobi, Kenya. 103 pages (see pp. 24–27). GIS database available online: http://www.grid.unep.ch/data/download/gnv174.zip. Accessed 28 Jun 2010

El Gamri T (2004) Prospects and constraints of desert agriculture: lessons from west Omdurman. Environ Monit Assess 99:57–73

El Gamri T, Saeed AB, Abdalla AK (2009) Rainfall of the Sudan: characteristics and prediction. Arts J 27: 18–35. Journal of the Faculty of Arts, Univ of Khartoum, Sudan. http://adabjournal.uofk.edu/current%20issue/ISSUES%20ENGLISH/El%20Gamri_%20Amir%2. Accessed 24 Feb 2013

El Tom MA (1975) The rains of the Sudan: Mechanisms and distribution. Khartoum University Press, Khartoum

Elagib NA, Elhag MM (2011) Major climate indicators of ongoing drought in Sudan. J Clim 409(3–4):612–625. doi:10.1016/j.jhydrol.2011.08.047

Fink AH, Vincent DG, Ermert V (2006) Rainfall types in the west African Sudanian zone during the summer monsoon 2002. Mon Weather Rev 134:2143–2164

Funk C, Brown M (2005) A maximum-to-minimum technique for making projections of NDVI in semi-arid Africa for food security early warning. Rem Sens Environ 101:249–256. http://earlywarning.usgs.gov/adds/pubs/ndvi_projections.pdf. Accessed 15 Jan 2013

GCCS (2004) Implementation plan for the global observing system for climate in support of the UNFCCC, Executive Summary, October 2004, GCOS – 92 (ES), (WMO/TD No. 1244). http://www.wmo.int/pages/prog/gcos/Publications/gcos-92_GIP_ES.pdf. Accessed 10 Oct 2012

GOSIC (2012) Global observing systems information center (GOSIC) is managed by NOAA as a data portal providing access to data and information identified by the global climate observing system (GCOS) and other member programs of the WMO and its collaborators. http://gosic.org/default.htm. Accessed 10 Oct 2012

Holden J (ed) (2005) An introduction to physical geography and the environment. ISBN 0131-2.761-5, Pearson Education Ltd, Edinburgh

Hulme Mike (1990) The changing rainfall resources of Sudan. Trans Inst Br Geogr NS 15(1):21–34

Hulme M (2001) Climate perspectives on Sahelian dessiccation: 1973–1998. Glob Environ Change 11(1):19–29. doi:10.1016/S0959-3780(00)00042-X

Jurkovica RS, Pasaric Z (2012) Spatial variability of annual precipitation using globally gridded data sets from 1951 to 2000. Int J Climatol. doi:10.1002/joc.3462

Kandji T, Verchot L, Mackensen J (2006) Climate change and variability in the Sahel region: Impacts and adaptation strategies in the agricultural sector. United Nations environment programme (UNEP), World Agroforestry Centre (ICRAF), Nairobi. http://worldagroforestrycentre.net/. Accessed: 4 Jul 2012

Kraus EB (1977) Subtropical droughts and cross-equatorial energy transports. Mon Weather Rev 105:1009–1018

Mahé G, Paturel J-E (2009) 1896–2006 Sahelian annual rainfall variability and runoff increase of Sahelian rivers. C R Geosci 341: 538–546

Ngigi SN (2009) Climate change adaptation strategies: water resources management options for smallholder farming systems in Sub-Saharan Africa. The MDG centre for east and southern Africa, The Earth institute at Columbia University, New York. ISBN 978-92-9059-264-8

Nicholson SE (1986) The spatial coherence of African rainfall anomalies: interhemispheric teleconnections. J Clim Appl Meteor 25:1365–1381

Nicholson SE (1989) Long term changes in African rainfall. Weather 44:46–56. doi:10.1002/j.1477-8696.1989.tb06977

Nicholson SE (2009) A revised picture of the structure of the monsoon and land ITCZ over West Africa. Clim Dyn 32(7–8):1155–1171

Nicholson SE (2011) Dryland climatology. Cambridge University Press, Cambridge, New York

Nicholson SE, Grist JP (2001) A simple conceptual model for understanding rainfall variability in the West African Sahel on interannual and interdecadal time scales. Int J Climatol 2:1733–1757. doi:10.1002/joc.648

Nicholson SE, Selato JC (2000) The influence of La Nina on African rainfall. Int J Climatol 20(14):1761–1776

Riddle EE, Cook KH (2008) Abrupt rainfall transitions over the greater horn of Africa: Observations and regional model simulations. J Geophys Res 113(D15): D15109. doi:10.1029/2007JD009202

Spinage CA (2012) African ecology—benchmarks and historical perspectives. Springer, Heidelberg

Sulieman HM, Buchroithner MF, Elhag MM (2012) Use of local knowledge for assessing vegetation changes in the southern Gadarif region. Afr J Ecol 50(2):233–242

Suzuki T (2011) Seasonal variation of the ITCZ and its characteristics over central Africa. Theor and Appl Climatol 103(1–2):39–60. doi:10.1007/s00704-010-0276-9

Thompson BW (1965) The climate of Africa. Cambridge University Press, London

Toulmin C (2009) Climate change in Africa. Zed Books, London

Tucker CJ, Justice CO, Prince SD (1986) Monitoring the grasslands of the Sahel 1984–1985. Int J Remote Sens 7:1571–1581

Tukey JW (1977) Exploratory data analysis. Addison-Wesley Publ. Co., Reading

UNEP (2003) Sudan: post-conflict environmental assessment. Chap 3: Natural disasters and desertification. UNEP disasters and conflicts programme. http://postconflict.unep.ch/publications/sudan/ Accessed 17 Feb 2013

UNEP (2008) Africa: Atlas of our changing environment. UNEP division of early warning and assessment (DEWA), Nairobi. ISBN: 9789280728712. http://www.unep.org/dewa/africa/africaAtlas/ Accessed 8 Jul 2012

UNEP (2010) Africa Water Atlas. UNEP division of early warning and assessment (DEWA) Nairobi: ISBN: 9789280731101. http://www.na.unep.net/atlas/africaWater/downloads/africa_water_atlas.pdf. Accessed 8 Jul 2012

Verdin JP, Funk CC, Senay GB, Choularton R (2005) Climate science and famine early warning. Phil Trans Roy Soc B—Biol Sci 360(1,463):2,155–2,168. ftp://chg.geog.ucsb.edu/pub/pubs/RoyalSociety_2005.pdf. Accessed 15 Jan 2013

Chapter 2
Analysis of Long-Term (100-year) Patterns of Rainfall Variability

Abstract Precipitation data from century-long rainfall records for three standard WMO rain gauge stations in the East Sahel—Khartoum (KHA), Kassala (KAS), and Gedaref (GED)—show strong interannual variability of rainfall in space and time. Spatial variations in the temporal variability of annual totals are assessed using ranked quantile statistics and the interquartile dispersion (IQD). The northern driest station, KHA, has an IQD of 0.62, and the southern wettest station, GED, has an IQD of 0.26. The middle station, KAS, has an IQD of 0.44. Comparing time series of annual totals for the three stations reinforces the extent of the interannual variability, with little evidence for long-term regional trends in the data. The annual values are reduced to three common types of normalized metrics: (1) standardized precipitation indices (SPI); (2) each year's total as a percent of the expected annual value; and (3) each year's total as a percent departure from the expected annual value. Single-station and multistation composites are compared to published results from other regions of the Sahel. While there are subdued indications that the results from GED, KAS, and KHA are sometimes coherent among several of the stations, and occasionally coherent with trans-African aggregations reported by other workers, such signals are most often dominated—even obscured—by episodic transients that are temporally incoherent over spatial scales on the order of the distance between the three study sites used here. The consequences present grave challenges to predicting and monitoring the type of interannual variability of rainfall that presents the most immediate and greatest threat to the sustainability of local cultures and economies in the East Sahel.

2.1 Study Area in East Sahel

The locations of the World Meteorological Organization's (WMO's) long-term rain gauges used in this study are shown in Fig. 2.1, relative to the national borders of Sudan, Eritrea, and Ethiopia, and the border of the Sudan State of Gedaref, one of the major agricultural regions of Sudan. Also shown, highlighting the sharp

J. F. Hermance, *Historical Variability of Rainfall in the African East Sahel of Sudan*, SpringerBriefs in Earth Sciences, DOI: 10.1007/978-3-319-00575-1_2, © The Author(s) 2014

Fig. 2.1 Location of WMO station gauges (*circular symbols*) used in this study. Elevation contours at 200 m intervals. Regional distribution of annual rainfall from the 1951–1980 UNEP/DEWA-GRID database (Deichmann and Eklundh 1991)

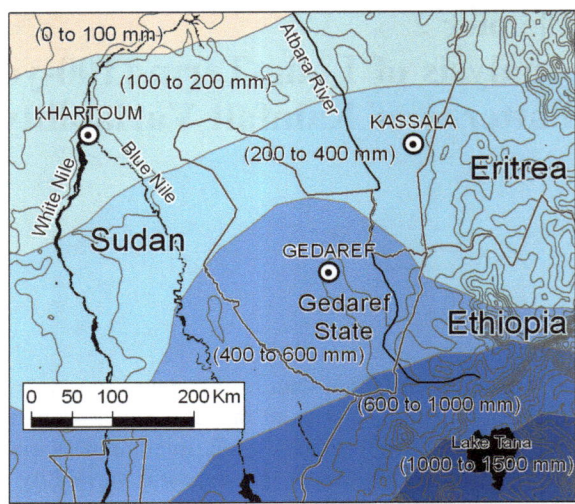

increase in the elevation of the Ethiopian Highlands to the southeast, are elevation contours at 200 m intervals.

Although not labeled, the respective contours may be referenced to the elevations of each of the three gauges used in this study given in Table 2.1. Also shown in Table 2.1 are the expected annual total rainfalls for each station based on several types of metrics. The first column under *Annual Rainfall Total (mm)* is the mean annual rainfall for each station from the 30-year period of record 1951–1980—the nominal time frame used by Deichmann and Eklundh (1991) for the construction of the GIS database plotted in Fig. 2.1. Next are the WMO Climate Normals from the 1961–1990 epoch. The third column lists the expected annual totals computed from the median of all annual totals from the entire period of record (100+ years) for the respective station.

The purpose of Table 2.1 is to demonstrate the type of differences one obtains by computing estimates of annual rainfall over different periods of record. For purposes of this report, unless explicitly noted otherwise, the practice will be to use the long-term annual total in the right-most column of Table 2.1—the expected value from the entire period of record (Table 1.1)—as the preferred metric for the expected annual rainfall.

2.2 Constructing Time Series of Annual Precipitation Totals

Figure 2.2 compares long term time series of annual totals for the three stations in the study in Fig. 2.1. Although the percent coverage of all three stations is 98 % or better (see Chap. 1, Table 1.1), there are, nevertheless, missing data. A primary

Table 2.1 Details on WMO station gauges

WMO station	Elev (m)	Period of record	Annual rainfall total (mm)		
			Mean annual (1951–1980)	WMO normal (1961–1990)[a]	Expected value from period of record[b]
Khartoum (KHA)	380	1899–2009	160	162	130
Kassala ((KAS)	500	1901–2009	291	251	280
Gedaref (GED)	599	1903–2009	592	603	600

[a] WMO Normals (2012). Annual mean of 30 years of annual values
[b] From Table 1.1, based on the median of 100+ years of data

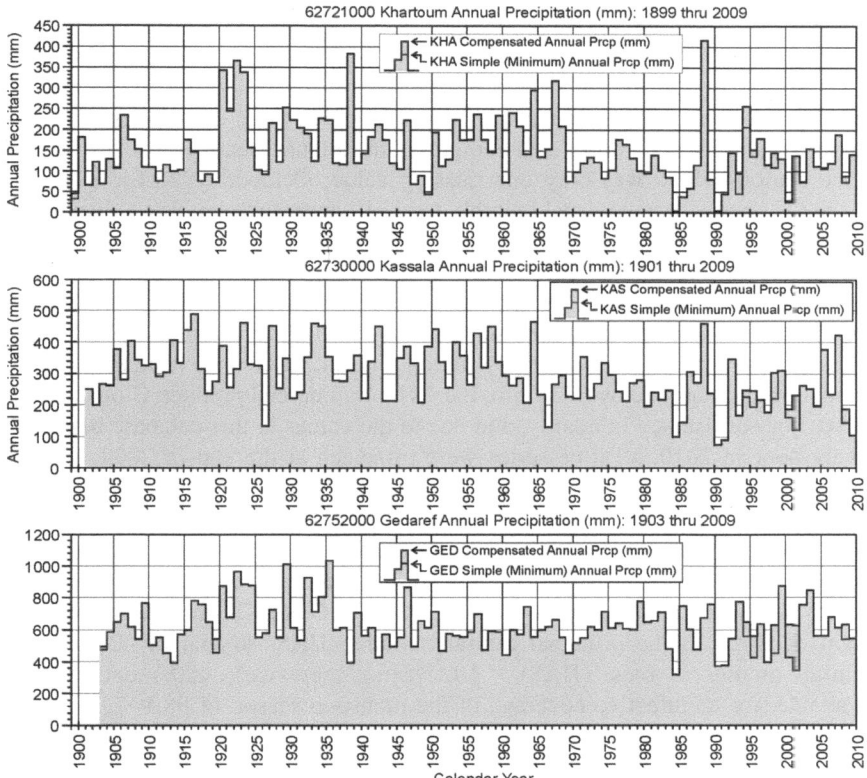

Fig. 2.2 Comparing the long-term time series of annual totals (mm) for Khartoum, Kassala, and Gedaref. See text for definitions of "simple" and "compensated" time series

concern, therefore, is to address the issue of these missing values. All of the annual totals reported here are synthesized from monthly values, so that over the 107 year period of record for Gedaref, for example, there is a potential for $107 \times 12 = 1,284$ monthly values, of which 17 values were flagged in the Global Historical Climate Network (GHCN) database as missing. However, a number of

Table 2.2 Missing principal rainfall months. (Boxes highlight particularly problem years.)

GED		KAS		KHA	
Years	Month	Years	Month	Years	Month
1919	6			1899	9
				1993	7
				1993	9
1994	6	1994	6	1994	8
1995	9	1995	9	1995	9
1998	7	1998	7	1998	7
2000	8	2000	6		
2001	6	2001	6		
2001	7	2001	7	2001	7
2008	9	2008	9	2008	9

missing values are from months where, in retrospect, there is typically little to no seasonal rainfall. For example, in computing the annual means in Table 2.1, of the three stations, there was only one missing value: October, 1973 for Khartoum. Since October has an expected monthly total of 1 mm, such missing values are not particularly important. Missing values in June (J), July (J), August (A), and September (S) are significantly more important.

When patching together annual totals from monthly values into a multiannual time series, missing months are handled in the following way. First, the potential impact of missing data is assessed by station. The data used were from the GHCN as formatted and supplied by EarthInfo, Inc. as its commercial product: Global Climate 2010. For consistency in quality, and due to the values in this database being relatively poor in 2010, all time series were truncated at the end of 2009. All trace precipitation values were set to 0 mm, and missing values for the respective periods of record were assessed, with the following results, and as tabulated in Table 2.2.

The GED period of record is from 1903 thru 2009 (107 years), for a total of 1,284 months. There were 20 missing values, leaving a total of 1,264 monthly observations, for an overall coverage of 98.4 %. Of the 20 missing values, only eight occurred in the principal rainfall months: JJAS, so that of 428 principal rainfall months (4 mos. (JJAS) × 107 years), there were 420 monthly values available, for an effective coverage of the monsoon season of 98.1 %.

The KAS period of record is from 1901 thru 2009 (109 years), for a total of 1,308 months. There were 14 missing values, leaving a total of 1,294 monthly observations, for an overall coverage of 98.9 %. Of the 14 missing values, only seven occurred in the principal rainfall months: JJAS, so that of four months (JJAS) × 109 year = 436 principal rainfall months, there are 429 monthly values, for an effective coverage of the monsoon season of 98.4 %.

The KHA period of record is from 1899 thru 2009 (111 years), for a total of 1,332 months. There were 24 missing values, leaving a total of 1,308 monthly observations, for an overall coverage of 98.2 %. There are three (JAS) principal rainfall months for KHA. Of the 24 missing values, only eight occurred in the principal rainfall months of JAS, so that of three monsoon months

(JAS) × 111 year = 333 principal rainfall months, there are 325 monthly values, for an effective coverage of the monsoon season of 97.6 %.

Table 2.2 provides details on the missing months during the principal rainfall seasons by year, which should be inspected to identify particular problem years where either several critical months were missing at a site (e.g. KHA for 1993), or several stations had missing months. Of all the years, only 1993 for KHA, and 2001 for GED and KAS, had more than one missing month per year. The years 1994, 1995, 1998, 2001, and 2008 had one missing month at all three stations.

Allowing that *some* information is generally of more help to an analysis than *no* information, these "missing months" are accounted for by substituting a statistical proxy—the monthly expectation value—determined from the 100+ long monthly time series for each respective station. For purposes of this report, to minimize the effect of the singular outliers evident in the raw observations, the expected value is computed as the median rainfall total for the respective calendar month computed from the entire period of record of observed values for the station. Typically, therefore, the expected value is estimated as the median of more than 100 annual samples for each month.

The results of this procedure are shown as the "compensated" annual precipitation in Fig. 2.2 where the step plot of the time series for the annual totals for the respective station bifurcates for years of missing data. Where double values appear for a year in Fig. 2.2, the lower curve is a result of assuming that no data equals zero data, and then simply zeroing out the monthly contribution to the annual total. The upper curve is for the case where if there are no data, then the value one might expect is the long-term monthly median, and the latter is substituted for the missing month. The annual total for the year when there are missing months, is the simple sum of the available data and the median value for the respective missing month(s). The resulting annual time series adjusted in this way are the "compensated time series" in Fig. 2.2. Other workers sometimes handle missing months by computing the annual total as the simple sum of all months for which values are available. The latter metric is referred to in Fig. 2.2 as the "simple" time series, which provides an estimate of the minimum value of the precipitation for the month. Statistically speaking, however, the expected precipitation for a month with missing data should be closer to the long-term median value, rather than the implied value of zero used in the simple sum. Caution, of course, needs to be exercised when drawing any conclusions from the results of either approach. For the remainder of this report, the three compensated time series in Fig. 2.2 will serve as the basic database of annual precipitation totals used in the analysis.

2.3 Spatial Differences in Annual Precipitation Totals

Overlaying the time series for the two stations Khartoum and Gedaref as in Fig. 2.3 emphasizes the spatial and temporal differences and similarities in annual precipitation over the span of the study area.

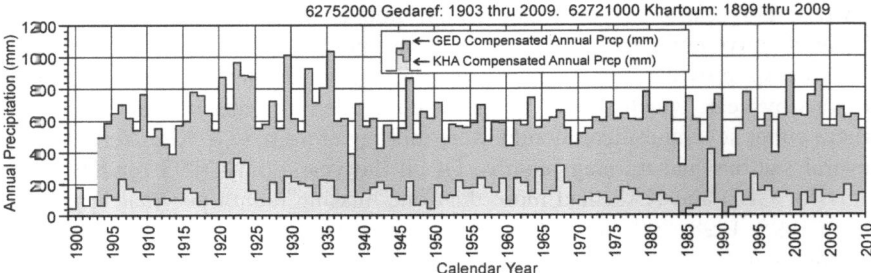

Fig. 2.3 Superposition of the compensated annual time series of monthly precipitation at Gedaref (GED; *dark grey*) and Khartoum (KHA; *light grey*). The label for the year is placed at the beginning of the respective year

The apparent intersection of the two curves in 1938 is due to the annual precipitation at Khartoum *increasing* in that year to a total of 383 mm (compared to an annual expected value of 130 mm), and the precipitation at Gedaref *falling* to 394 mm (compared to an annual expected total of 600 mm); the 1938 Gedaref value is too close to that of Khartoum to resolve the difference on this graph. The point is that annual totals in 1938 show a strong negative correlation of precipitation between the two stations over a north–south distance of 220 (actually 216) km, apparently an East Sahel counterpart to Nicholson's (2009) "dipole effect," a term she uses in the West Sahel to describe annual anomalies of the opposite sign north and south of an intermediate demarcation line. As in the West Sahel, other years for the East Sahel in Fig. 2.3 show a positive correlation—a positive anomaly in the north is associated with a positive anomaly in the south, or a negative anomaly in the north is associated with a negative anomaly in the south. Examples include 1920 thru 1923, 1946, 1960, 1967, 1969, 1984, and 1990, among others. Noteworthy are the extreme drought years 1984 and 1990 on the two time series. Also noteworthy because of the *lack* of significant correlation is the extreme flood year of 1988 at Khartoum (Sutcliffe et al. 1989; Hulme and Trilsbach 1989), which shows up dramatically in the annual total at the station, but lacks a significant counterpart at Gedaref to the south, underscoring the spatial variability of annual rainfall in the East Sahel. The high rainfall year in 1999 at Gedaref lacks a counterpart at Khartoum.

To better compare common patterns of coherent and incoherent climate signals among the three reference stations, the following section will identify the long-term baseline for each station, and quantitatively characterize the expected range of interannual variations from each station's baseline. Following that, several alternative procedures will be used to normalize the amplitudes of interannual variations so as to enhance the signal against the static background imposed by regional precipitation gradients.

2.4 Expected Range of Annual Precipitation Totals

One measure of the interannual variability of rainfall in the East Sahel is to compare the statistical distributions of the annual precipitation totals recorded by each station. The stepped line graph (histogram) in each plot in Figs. 2.4, 2.5, and 2.6 is for the binned class density (left axis) presented as the probability density per bin width. For example, in the case of Gedaref (Fig. 2.4) there is approximately a 32 % probability that the measured total rainfall for a year will fall in the 100 mm wide bin between 500 and 600 mm.

The more continuous line with symbols is the cumulative probability density (right axis), which is the probability that the annual precipitation is equal to or less than the corresponding value along the abscissa. In Fig. 2.4, as shown by the arrows, there is a 50 % cumulative probability that annual rainfall will be equal to or less than 600 mm.

The tables corresponding to each plot are numerical summaries of the information presented in the graph. These parameters include the *first*, *second*, and *third* *quartiles* (Q25, Median (or Q50) and Q75), respectively. A parameter that is

Fig. 2.4 Statistics on annual precipitation totals for GED (see summary in Table 2.3)

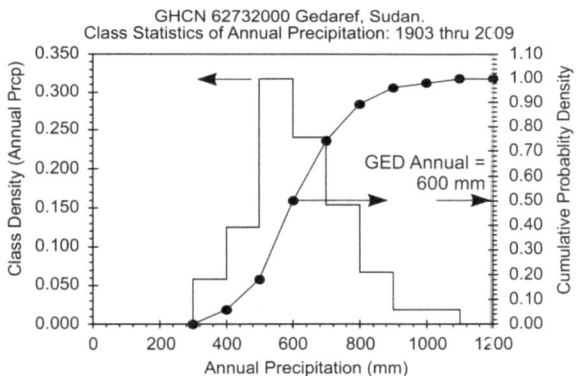

Fig. 2.5 Statistics on annual precipitation totals for KAS (see summary in Table 2.4)

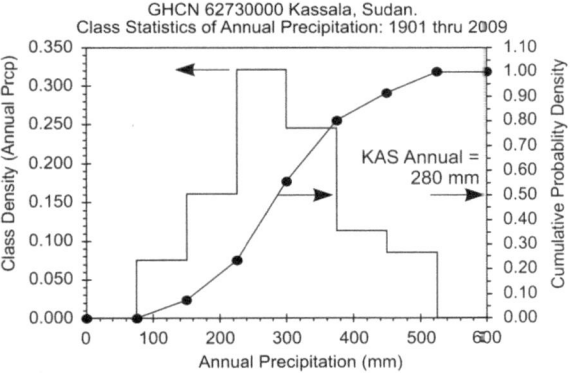

Fig. 2.6 Statistics on annual precipitation totals for KHA (see summary in Table 2.5)

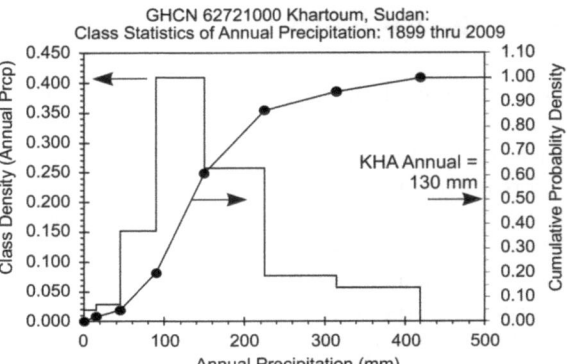

derived from the Q75 and Q25 values, is the *interquartile range* (IQR), defined here and used elsewhere in this report as IQR = (Q75−Q25).

Often it is useful to know the *relative* variability of rainfall at a site, rather than just the IQR, which is a measure of the absolute range. When dealing with normally distributed (or Gaussian-type) variables, one might use the *dispersion*, defined as the standard deviation (*sd*) divided by the mean value (μ) (so that the *normal dispersion* = sd/μ). When dealing with datasets having significant outliers, it is advisable to use quantile statistics, so that the preferred metric to be used in this report for relative variability is the *interquartile dispersion* (IQD), which is the interquartile range normalized (divided) by the median, such that IQD = IQR/Q50, which becomes IQD = (Q75−Q25)/Q50. Also given in Tables 2.3, 2.4 and 2.5 are the extreme maximum (Max) and the extreme minimum (Min) annual totals for all years of record, respectively, followed by the extreme range, which is the difference between the maximum and minimum annual totals of all years for the respective station. Next, the count or number of years is given for which observations are recorded, which (accounting for missing data) is less than or equal to the number of years of record. For example, the period of record for Gedaref is 1903–2009, or 107 years of record, but there are only 104 years for which totals are available. Finally, the bin width used to generate the respective histogram in the figure is given. Again, for example, the bin width used for the Gedaref class density of annual precipitation is 100 mm—thus, from the class density plot in Fig. 2.4 (according to the left ordinate), there is approximately a 0.25 (25 %) probability that the total annual rainfall for Gedaref will be greater than 600 mm and less than 700 mm. This implies a return period for this latter range of approximately four years, which is to say that one out of every four years is expected to have an annual total between 600 and 700 mm.

2.4.1 Gedaref (Median Annual Prcp = 600 mm)

According to the cumulative probability density in Fig. 2.4 (right vertical axis), there is a 0.50 (50 %) probability that the annual total will be equal to or less than

Table 2.3 Annual
precipitation for GHCN
62752000, Gedaref
(1903–2009)

Q25 = 548 mm
Median (Q50) = 600 mm
Q75 = 702 mm
IQ range (IQR) = 154 mm
IQ dispersion (IQD) = 0.26
Max = 1035 mm
Min = 321 mm
Extreme range (mm) = 714 mm
Count = 104 years
Bin width: 100 mm

600 mm. The cumulative probability that the annual total will be less than or equal to 700 mm is between 70 and 80 %; actually, from the third quartile (Q75) in Table 2.3, there is a 0.75 (75 %) probability that the annual total will be less than or equal to 702 mm. The IQD for Gedaref is approximately 26 % (rounded from 0.256), which is typical for precipitation data from other stations in the Sahel having similar expected totals.

2.4.2 Kassala (Median Annual Prcp = 280 mm)

Based on the classification and cumulative probability plots (Fig. 2.5), and the quantile statistics in Table 2.4, the annual total rainfall for Kassala (based on the Median (Q50) = 280 mm) is expected to be approximately half that for Gedaref. Kassala has an interquartile range of IQR = 122 mm, which leads to a value for the interquartile dispersion of IQD = 44 % (rounded from 0.435), implying a significantly larger *relative* interannual variability at KAS than at GED. Note, however, that the interquartile range (IQR) at KAS = 122 mm, which is *smaller* than the interquartile range (IQR) at GED = 154 mm. One needs to be careful to discriminate between *relative* variability—in the sense of *normalized* variability (e.g. IQD)—and *absolute* variability (e.g. IQR), which involves a simple difference between the upper range Q75 and the lower range Q25 metrics.

Table 2.4 Annual
precipitation for GHCN
62730000 Kassala
(1901–2009)

Q25 = 229 mm
Median (Q50) = 280 mm
Q75 = 351 mm
IQ range (IQR) = 122 mm
IQ dispersion (IQD) = 0.44
Max = 488 mm
Min = 76 mm
Extreme range (mm) = 412 mm
Count = 106 years
Bin width: 75 mm

Table 2.5 Annual precipitation for GHCN 6272 1000 Khartoum (1899–2009)	Q25 = 101 mm Median (Q50) = 130 mm Q75 = 182 mm IQ range (IQR) = 81 mm IQ dispersion (IQD) = 0.62 Max = 415 mm Min = 4 mm Extreme range (mm) = 411 mm Count = 105 years Bin edge(s): 0, 15, 45, 90, 150, 225, 315, 420, 540 mm

2.4.3 Khartoum (Median Annual Prcp = 130 mm)

For the case of Khartoum in Fig. 2.6 and Table 2.5, the relative interannual variability is greater than the other two stations. For KHA, the IQD = 62 % (rounded from 0.623), compared to 44 % for KAS and 26 % for GED. Moreover, in plotting the statistics, the width of the bin is not constant, as in the previous cases for GED and KAS, but increases *geometrically* with increasing precipitation total. Note that the binning information is in terms of the *edges* for Khartoum, rather than the *width* as for Gedaref and Kassala, suggesting a log normal distribution for annual precipitation. The *lower* interquartile (Q50–Q25) = 29 mm, whereas the *upper* interquartile (Q75–Q50) = 52 mm is almost twice as large, underscoring the asymmetry of the distribution. This type of asymmetry is responsible for biasing the annual mean to higher values than the median. The median is preferred as the best estimate for the expected annual total, since 50 % of all the annual values over the 100+ year period of record are above this value, and 50 % are below (Table 2.5).

2.5 Normalized Precipitation Metrics to Describe Interannual Variability

A principal objective of climate analysis is to demodulate coherent from incoherent temporal climate signals common to a particular region. For example, the objective might be to identify long-term trends in, or multi-year modulations of, annual rainfall on a broad, sub-continental scale (e.g. Nicholson 1986, 1989). To do so usually requires a metric that minimizes the "noise" from singularly local rainfall events that appear on isolated station records, some of which are evident in the annual totals in Fig. 2.2. A common practice among data analysts is to *stack* or *average year-by-year* time series from a number of adjacent stations, so that the contributions from individual storm events spatially isolated to a particular station tend to be filtered out.

However, since simple multi-station annual means will be biased by the stations with the largest annual totals (a particular problem in regions with large spatial gradients in annual totals as in the Sahel), it is appropriate to rescale, or normalize, single-station observations prior to aggregating (or stacking) the parameters into a composite metric (e.g. Kraus 1977). In some cases, it might be advantageous to subtract the expected value of each station's annual rainfall, prior to rescaling and stacking. These regional composites have been quite useful in assessing the ability of computer-generated models of atmospheric circulation to reproduce some of the large-scale regional climate patterns discussed below (Wang et al. 2004; Biasutti et al. 2008).

However, a concurrent application of normalization applies to single-station data as well, for example to allow the analyst to compare the temporal patterns of precipitation among individual stations in a cluster, where the annual totals at each may be significantly different, as is the case for the three reference stations in the East Sahel. In other words, it may be useful to compare the *relative* temporal variations for each site, without being concerned about the absolute annual totals. This section reviews three methods commonly used to provide such normalized values.

2.5.1 The Standardized Precipitation Index

Single Station and Multi-Station Composite SPIs One of the most common metrics for representing the long term behavior of precipitation and droughts as time series is to first standardize the station observations as discussed below, and then, if one needs to minimize local interstation variability, to employ a multi-station composite of the standardized variate (see, for example, Kraus 1977; Nicholson 1986; Agnew 2000; Hulme 1991, 2001; Bell and Lamb 2006). The resulting parameter, referred to by many workers and in this report as the *standardized precipitation index* (SPI), is also sometimes referred to in the literature as the *drought index*, the *standardized precipitation anomaly* (SPA), the *normalized rainfall departure* (Bell and Lamb 2006), or—for the case of annual data—Nicholson (1986) refers to the parameter as the *annual rainfall departure*, which she shows, when composited among clusters of stations from similar precipitation environments in Africa, reveals a remarkable temporal coherence from the Southern Kalahari to the Sahel.

As the name suggests, the SPI applies the concept of a *standardized variate* from statistics to a time series of precipitation values. The SPI for a site is usually calculated at monthly or annual intervals, and is defined in the following way for a time series of n samples. For the ith member of the time series, the SPI value is

$$SPI_i = \frac{P_i - P_{mean}}{sd_P} \qquad (2.1)$$

where P_i is the value of the ith sample of the time series, P_{mean} is the mean value of the n members of the time series (or, in some cases, a reference subset of the n total samples), and sd_P is the standard deviation of the n samples from the mean P_{mean}.

If the objective is to composite SPI values from different stations, the SPI for the jth station is defined as

$$SPI_{ij} = \frac{P_{ij} - P_{mean_j}}{sd_{P_j}} \qquad (2.2)$$

Often, these single-station time series are useful in their own right for comparing inter-station similarities and differences. However, to combine the SPIs from a number of stations into a single composite time sample, a mean SPI is defined as

$$\langle SPI \rangle_i = \frac{1}{m} \sum_{j=1}^{j=m} \left[\left(P_{ij} - P_{mean_j} \right) \Big/ sd_{P_j} \right] \qquad (2.3)$$

where m is the number of stations, and $\langle SPI \rangle_i$ is the composited SPI value (or mean value of SPI) for the m stations at the ith time sample.

Kraus (1977) argues that such a standardized variable, when composited among a number of stations, tends to compensate for the effect that stations with smaller annual totals typically have larger variability. Nicholson (1986) asserts that the standardized variable is an effective way to merge mixed datasets having diverse statistics. After computing the SPI time series, some workers then smooth the data—such as by multi-year running averages—to obtain better views of the long-term, multi-year behavior of precipitation in a region, although this report will use un-smoothed annual values in order to better identify subtle short-term differences among stations.

Multi-Station, Composite SPIs for the African Sahel One of the earliest trans-African SPI constructions to indicate what would turn out to be the beginning of the great Sahel drought of the 1970s and 1980s was produced by Kraus (1977), who composited all data available at the time from African rain gauge stations between 30° West and 60° East, and 10° North to 25° North. A plot of his tabulated values is shown in Fig. 2.7. There were approximately 10–15 stations available for the first several decades, increasing to 25–30 stations for the last several decades of the study (see Kraus 1977, for details). The well-known, long-term decline in annual precipitation across Africa from the 1950–1960 "wet"

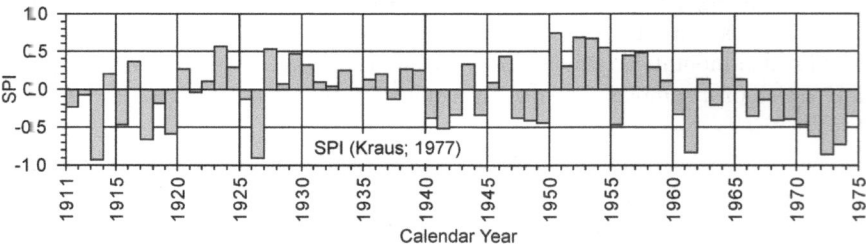

Fig. 2.7 Trans-African composite standardized precipitation indices compiled by Kraus (1977), plotted by author. The annual label is shown at the beginning of the respective year

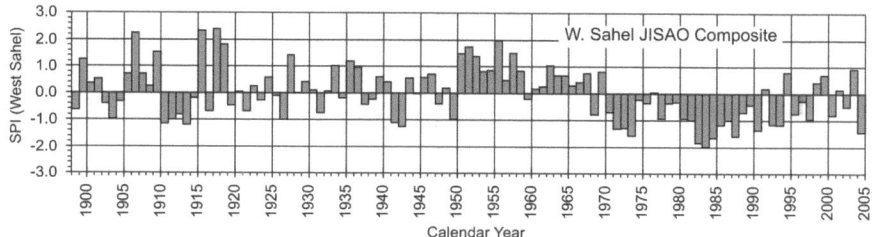

Fig. 2.8 West Sahel JISAO SPI reference database for 1898–2004 (JISAO 2005). The annual label is shown at the beginning of the respective year

years to the "dry" years of the 1970s is quite marked—in fact, there is evidence for several precursory dry periods throughout the 1960s.

A composite SPI time series for an array of gauges in the West Sahel for the period 1898–2004 (JISAO 2012) is shown in Fig. 2.8. The latter was constructed by the Joint Institute for the Science of the Atmosphere and Oceans (JISAO) at the University of Washington, building on the earlier database assembled by Nicholson (1979). The standardized, multi-station database in Fig. 2.8 will be referred to in this report as the W. Sahel JISAO SPI.

Although there are substantial differences in detail, the broad-scale form of the W. Sahel JISAO SPI composite in Fig. 2.8 compares favorably to that of Kraus (1977) in Fig. 2.7, in particular the persistent positive ("wet") anomaly in the early 1950s, and the deep negative ("dry") anomaly in the 1970s thru the 1980s. These broad-scale features appear in other standardized rainfall indices in the literature as well, particularly to the standardized data of Nicholson (1986) for the period of record 1901–1975, and Nicholson (1994) for the period 1901–1994; to the SPI of Agnew (2000) for the period 1931–1990; to the normalized annual rainfall anomalies of Wang et al. (2004); to the average normalized departure of Bell and Lamb (2006) for the epoch 1941–2004; to the SPI of Ali and Lebel (2008) for 1901–2006; among others.

Some of the actual amplitudes and the signs of the smaller anomalies may differ among these datasets somewhat, due to different time bases used to calculate the reference means and standard deviations. For example, the West Sahel JISAO reference data in Fig. 2.8 are standardized to the average for the complete period of record, 1898–2004, for each station, prior to compositing.

Some version of multi-station, composite SPIs has provided the basis for identifying many of the principal large-scale, interannual, and interdecadal precipitation events over the last century of climate observations. For example, reaffirming and adding to the 1911–1974 SPI analysis of Kraus (1977) shown in Fig. 2.7, Nicholson (1986) used the SPI in a study of the synchronicity of climate extremes across the African Sahel, pointing out the dry or drought conditions in the 1910s, the 1940s, and after the 1960s, with the 1950s being abnormally wet. Her standardized time series ending in 1975 catches the beginning of what became the decadal-long Sahelian drought period (later corroborated by the extended SPI time series of Hulme 2001), an event some feel was the largest regional climate change experienced on earth during the last half-century (Bell and Lamb 2006). According to such regional precipitation indicators, the drought lasted throughout the 1980s, followed by a modest rainfall recovery, punctuated by a number of very dry years, to the present time.

Single-Station SPIs for the East Sahel As mentioned previously in this report, local, single-station SPIs have their use as well. Figure 2.9 compares the single-station SPI time series for the three reference stations in the East Sahel using the compensated annual totals in Fig. 2.2. In Fig. 2.9, the SPI time series are compared (a) to each other, (b) to a composite (mean) of all three series, and (c) in the bottom panel, to the standard West Sahel JISAO SPI reference database in Fig. 2.8.

Thus, Fig. 2.9 (with Fig. 2.7) compares the local precipitation patterns from the three study sites used in this report against the backdrop of regional precipitation for the larger area of the African Sahel. The top-three time series in Fig. 2.9 are individual station SPIs for the respective station in the East Sahel: KHA, KAS, and GED. The fourth panel from the top is the composite SPI for these three stations. The bottom panel reproduces the West Sahel JISAO SPI from Fig. 2.8 for the period after 1903 synchronous with the periods of record for the three local stations. Comparing several features of the East Sahel three-station composite with the West Sahel JISAO SPI, the deep drought period of the 1970s and 1980s in the West Sahel is not as well developed for the three reference stations in the East Sahel. Certainly the anomalously wet summer of 1988 in the East Sahel is a discordant event in the overall rainfall pattern. However, there is an anomalous period of increased precipitation in the East Sahel from 1920 through 1935 that does not have a significant counterpart in the West Sahel, but does have a counterpart in the trans-African SPI analysis of Kraus (1977) in Fig. 2.7. Thus, the trans-African SPI of Kraus (1977) in the 1920–1935 epoch appears to be tilted toward the three-station composite SPI in the East Sahel (Fig. 2.9), having little similarity to the West Sahel SPI (Fig. 2.8). The opposite seems to be the case for the deep drought of the 1970s, where the trans-African SPI is similar to the West African SPI.

2.5.2 Percent of Expected Annual Precipitation

An alternative index to highlight the relative variability of rainfall is to normalize the annually observed rainfall by the long-term expected value for the annual total at the station. Each normalized annual value then becomes the percent of the expected annual total for the respective year. This parameter is particularly useful to agriculturists and local planners, allowing one to track the true variability of rainfall at a single station, while comparing relative changes at stations having grossly different annual totals. Here, this metric is computed as the respective annual total for the year (compensated for missing months, as in Fig. 2.2) divided by the median of annual totals for the complete period of record for each station as given in Table 2.1. Note that whereas some authors prefer to normalize by the

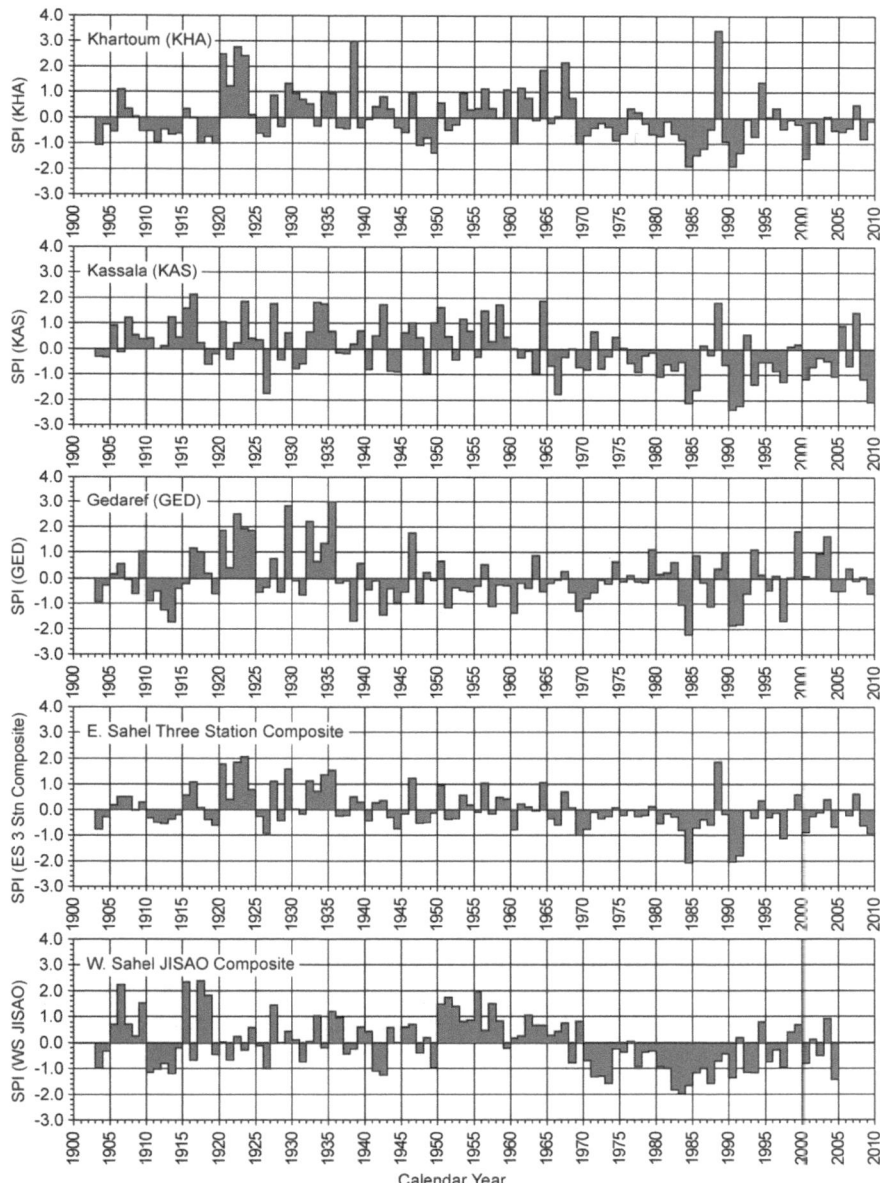

Fig. 2.9 Standardized precipitation indices (SPIs) from the Sahel. The top-three time series are individual station SPIs for the three study stations: KHA, KAS, and GED in the East Sahel. The fourth panel from the *top* is the composite (simple mean) of the three study stations. The *bottom* panel is a regional composite of SPIs from the West Sahel published by the Univ. of Washington (JISAO 2005). Compilation of the latter product was discontinued in 2005

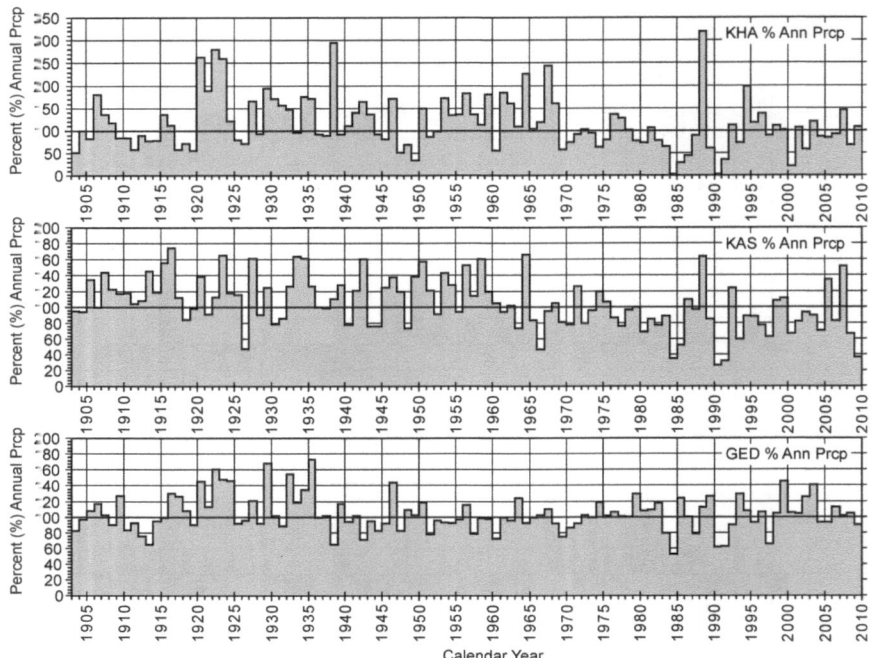

Fig. 2.10 Annual percent of expected precipitation for KHA, KAS, and GED. The 100 % level for each station's time series is highlighted in the foreground by the continuous horizontal line. Note that the plots for GED and KAS axes have common ranges of 200 %, but that, because of the significantly greater relative variability for KHA, the range of its plot extends to 350 %

long-term annual mean, the mean tends to be biased due to the type of outliers in the Sahelian rainfall data, and not so representative of the actual expected value. Here, the median is a more robust estimator of the expected value, and the results of the normalization are shown in Fig. 2.10.

Not surprising is the strong interannual variability of precipitation totals. The median of the absolute deviations (MAD) of the annual percent of expected precipitation is 10 % for GED, 20 % for KAS, and 35 % for KHA, a pattern consistent with an alternative metric of variability, the interquartile dispersion, in Table 2.1, where IQD(GED) = 26 %, IQD(KAS) = 44 %, and IQD(KHA) = 62 %. The scale is different for the two metrics, but the relative proportions of interannual variability among the three stations are roughly similar, and proportional to the relative differences in annual rainfall among the stations—the latter is largely a consequence of the normalization used.

Extreme events are significantly more extreme in relative terms for Khartoum, along the northern edge of the Sahel. Figure 2.10 emphasizes that rainfall at Khartoum during the extreme wet year of 1988 increased to 320 % of normal precipitation, whereas during the widespread drought year of 1984, precipitation dropped to 3 % of its expected annual value. Noteworthy is the wet period at Khartoum and Gedaref during the period of 1920–1925, but which is not particularly evident in the Kassala data. However, the higher rainfall at Gedaref for the four years 1933–1936 has a counterpart in the Kassala record, with only a two-year (1934–1935) concordance with data from Khartoum.

Several significant aspects of multiannual modulations of the precipitation patterns can be identified in Fig. 2.10 for each site. A case might be made that the Khartoum (KHA) sector of the Sahel—and perhaps Gedaref—emerged from the 19th century with a slightly lower-than-normal precipitation, where in 1920 the region entered a wet period, and Khartoum in particular experienced a 50-year period (1920–1970) of slightly higher-than-normal rainfall. For the last 40 years (after 1970), Khartoum and Kassala have experienced a somewhat lower rainfall, with of course the great Sahel drought of the 1970s and 1980s having a significant overprint on the climate.

The long-term pattern of precipitation at Kassala (KAS) suggests that the early decades of the period (1903–1965) had characteristically higher-than-normal rainfall, which was followed by a relatively short transition to somewhat lower-than-normal rainfall that lasted until the end of record in 2009. According to Fig. 2.10, long-term fluctuations of precipitation at Gedaref (GED) are much more benign in amplitude. The great Sahel drought of the 1970s and 1980s had a much weaker overprint on annual totals at Gedaref (GED), certainly much less than the 30-year dry period seen elsewhere in the Sahel described by Nicholson and Grist (2001). Moreover, there is a suggestion of a slight enhancement of precipitation at Gedaref during the two decades from 1916 thru 1935, where the expected annual total for the period is 120 % of the long-term (106 yr) expected value.

2.5.3 Percent Departure from Expected Annual Total

A third normalized metric is the percent departure ($\Delta P(\%)$) from the expected value of the annual total rainfall for the respective station (Kraus 1977; Nicholson 1989). For a particular station j, and for a particular year i, one computes

$$\Delta P(\%)_{ij} = \frac{P_{ij} - \langle P_j \rangle}{\langle P_j \rangle} \tag{2.4}$$

where $\langle P_j \rangle$ is the expected value for annual rainfall for the jth station, and F_{ij} is the annual total for the ith year at the jth station. For regional studies, the respective annual value $\Delta P(\%)_i$ can be composited from a number of stations according to

$$\Delta P(\%)_i = \frac{1}{m} \sum_{j=1}^{j=m} \left[(P_{ij} - \langle P_j \rangle) / \langle P_j \rangle \right] \tag{2.5}$$

where m is the number of stations, and $\langle P_j \rangle$ is the respective expected value of the annual total for the jth station. Both Kraus (1977) and Nicholson (1989) define this

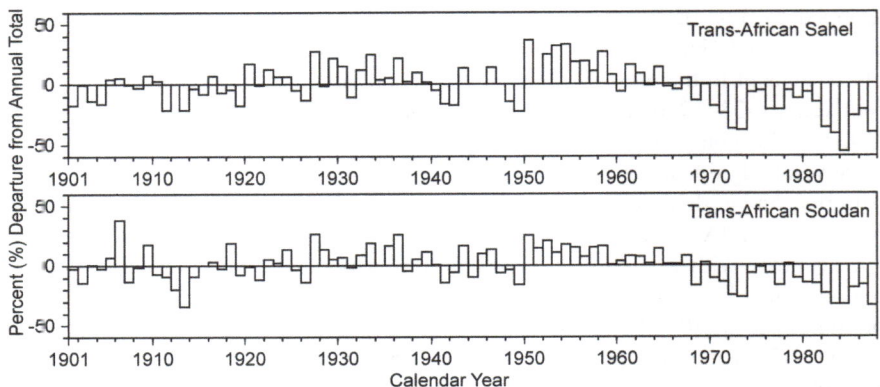

Fig. 2.11 Composited time series of the percent departure from annual mean for the period 1901 thru 1987, from two sub-parallel precipitation zones spanning Africa south of the Sahara (after Nicholson 1989). (Note tic marks at two-year intervals; annual label at beginning of the year.)

parameter as the percent departure from the long-term mean. Two examples from Nicholson (1989) of such composite time series are shown in Fig. 2.11, which compares the rainfall history from 1901 thru 1987 for the two sub-parallel semi-arid precipitation zones transecting Africa south of the Sahara shown in Fig. 2.12—defined here, following Nicholson (1989) as the Trans-African Sahel and the Trans-African Soudan.

Nicholson (1989) refers to the northern-most zone, respectively, as the "Sahel" (where the range of annual rainfall is 100–400 mm) and the southern-most zone as the "Soudan" (where the range of annual rainfall is 400–1200 mm). This classification differs somewhat from the way that the "Sahel" is generically defined in this report: the region of the south of the Sahara having a total annual rainfall of between 100 and 600 mm (see Fig. 2.1). The Sahel as defined in the current report (100–600 mm) straddles portions of Nicholson's (1989) two trans-African zones shown in Fig. 2.12.

The remarkable aspect of the two series in Fig. 2.11, as Nicholson (1989) points out, is the striking coherence of rainfall behaviors across Africa. Both time series

Fig. 2.12 Two of Nicholson's (1989) precipitation zones overlaid on a map produced from USGS NDVI data for July 1992

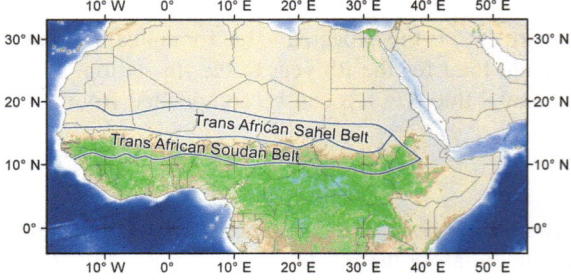

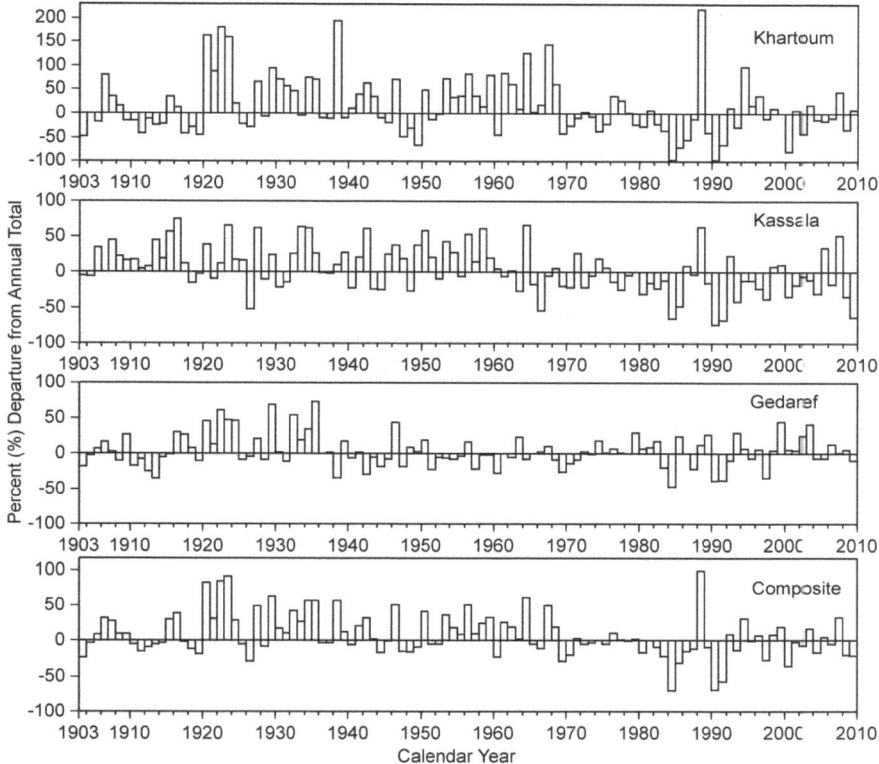

Fig. 2.13 Percent departures from the annual expected total (1903–2009) for the three study sites in the East Sahel. The composite in the bottom panel is the simple mean of the individual time series shown. (Note tic marks are at two-year intervals.)

show subdued values at the beginning of the 20th century, with an increase in yearly totals in the late 1920s to the late 1930s, followed by a period of normal rainfall from the later 1930s to 1950, then another period of increased precipitation from 1950 to the mid 1960s. The latter period of higher-than-typical precipitation is followed by a general decline in rainfall into the well-documented Sahel desiccation of the 1970s thru the 1980s.

The time series in Fig. 2.2 for the three station annual totals for 1903 thru 2009 have been synthesized into percent departures from their respective expected annual value in Fig. 2.13. For reasons already given, data from the three stations in the East Sahel have been referenced to the more robust, median of the annual totals from each of the respective 100+ year periods of record.

Several features in Fig. 2.13 for data from the three study sites in the East Sahel are seen to be similar to those for the trans-African data of Nicholson (1989) in Fig. 2.11. In particular, with reference to the three-station composite results in the bottom-most

Table 2.6 Summary of departure of annual rainfall from expected total (mm): 1984–1994

Station	Departure from expected annual (mm)		
	KHA	KAS	GED
Mean	−46	−81	−38
Q25	−90	−169	−198
Q50	−61	−78	−27
Q75	−19	−14	120
Minimum	−126	−204	−279
IQR	−71	−155	−318

The anomalous year 1988 is rejected

panel of Fig. 2.13, the 20th century begins with a generally normal rainfall, followed in the 1920s and 1930s by a wet period, which appears to be more locally enhanced in the East Sahel (Fig. 2.13). Annual rainfall in both figures returns to a somewhat unremarkable, normal period from 1940 to the early 1950s, followed by another wet period lasting into the 1960s, then by the profound drought of the 1980s.

Although the year-by-year coherence between the composited trans-African time series (Fig. 2.11) and the multi-station composite of the East Sahel time series (Fig. 2.13; bottom panel) is not particularly striking, there is a common pattern in that both the trans-African data and the individual station data in the East Sahel indicate that—in relative terms—the northern Sahel was significantly more affected by the drought of the 1980s than the southern Sahel (Nicholson 1989). This is most striking by comparing the relative departures between the north and the south over the 1984–1994 time frame. Compare, for example, the composite Trans-Sahel versus the Trans-Soudan time series of Nicholson in Fig. 2.11, and compare the normalized time series for Khartoum versus Gedaref in Fig. 2.13. The percent departures are significantly more negative in the north than the south Sahel, and the latitudinal gradient in variability is readily seen in Fig. 2.13 over distances of 100–200 km, by comparing data from the three study sites.

Table 2.6 presents the interstation variabilities during the drought of the 1980s from the perspective of un-normalized metrics. The statistical values are based simply on the actual annual departures (in mm) during the years 1984–1994, having dropped values for the highly anomalous rainfall year of 1988. Ten annual values (rejecting 1988) were used from each station to compute the statistics in Table 2.6. Recall that the long-term expected annual total for each of the stations is: 130, 280, and 600 mm for KHA, KAS, and GED, respectively. In terms of real—not relative—differences, Table 2.6 shows that the overall loss of rainfall during this decade for these three stations was actually quite modest, from the point of view of either the median or the mean of the departures (say 40–80 mm, overall). However, the extremes of the departures are a different story; and the spatial pattern of variability is just the opposite of the normalized metrics shown in Fig. 2.13. The most anomalous station is Gedaref, the southernmost site, for the lower quartile (Q25 = −198 mm) and for the greatest annual loss—or most negative departure (−279 mm)—of record (1984–1994). The lower quartile implies that 25 % of the years (one out of four) at Gedaref had 198 mm less rainfall than normal. The lower quartile for Khartoum for 1984–1994 is −90 mm. Thus, the actual rainfall loss at Gedaref for the decade is twice that at Khartoum.

2.6 Variability and Synchrony of Annual Rainfall in Space and Time

Comparing the GED, KAS and KHA time series in Figs. 2.2, 2.9, 2.10 and 2.13 shows that there are years (1964, 1988, 1984, 1991)—even multiple years (such as 1920–1925, 1991–1992)—when the sign and intensity of rainfall is coherent between two of, or all three of the three stations. Moreover, there are years where rainfall totals are quite incoherent among the three stations in the East Sahel. In the background, although often dominated by sharp episodic spatial and temporal incoherent transients, are decadal-long patterns that are grossly similar among the three-station time series, and between the three-station composites and composite time series from across Africa. As others have argued, this common background pattern implies underlying regional climate processes driven by common climate forcing mechanisms in the global oceans and atmosphere (Hulme 2001; Giannini et al. 2003; Wang et al. 2004; Zhang and Delworth 2006; Biasutti et al. 2008). However, the individual annual time series for the East Sahel considered in this chapter show that it is seldom that the regional signatures of global teleconnections are not dominated by local interannual anomalies. The focus throughout this report will be not so much on the similarities, but on the often not-so-subtle *differences* between the various climatic phases of the respective time series. Thus, for example, while according to the composite metrics in Figs. 2.7, 2.8, and 2.11, the climate of the trans-African Sahel was undergoing a regional multi-decadal transition from a "wet" regime in the 1950s and 1960s, to the "dry" regime throughout the 1970s and 1980s, local rainfall patterns at the three reference stations in the East Sahel (Figs. 2.2, 2.9, 2.10, and 2.13) fluctuated interannually among each other in sign, and by several hundred percent in rainfall totals. This chapter has emphasized the interannual temporal variability of rainfall at single stations, and the attendant spatial variability among several stations. Later chapters will analyze this variability at finer, interseasonal, and intraseasonal time scales, and finally this report will explore the variabilities in duration, frequency, and intensity among discrete storm events, and the consequent interstorm dry periods.

References

Agnew CT (2000) Using the SPI to Identify Drought. Drought Network News (1994–2001). Paper 1: http://digitalcommons.unl.edu/droughtnetnews/1. Accessed 29 Apr 2012

Ali A, Lebel T (2008) The Sahelian standardized rainfall index revisited. Int J Climatol Published online in Wiley Inter Science, (www.interscience.wiley.com) DOI: 10.1002/joc.1832

Balme M, Vischel T, Lebel T, Peugeot C, Galle S (2006) Sahelian water balance: impact of the mesoscale rainfall variability on runoff. Part 1: rainfall variability analysis. J Hydrol 33:336–348

Bell MA, Lamb PJ (2006) Integration of weather system variability to multidecadal regional climate change: The West African Sudan-Sahel zone, 1951–1998. J. Climate 19(20):5343–5365

Biasutti M, Held IM, Sobel AH, Giannini A (2008) SST forcings and Sahel rainfall Variability in simulations of the twentieth and twenty-first centuries. J Clim 21:3471–3486

Deichmann U, Eklundh L (1991) Global digital data sets for land degradation studies: a GIS approach. GRID Case Study Series No. 4; UNEP/GEMS and GRID; Nairobi, Kenya; 103 pages (see pp. 24–27). GIS database available online: http://www.grid.unep.ch/data/download/gnv174.zip. Accessed 28 Jun 2010

Giannini A, Saravanan R, Chang P (2003) Oceanic forcing of Sahel rainfall on interannual to interdecadal time scales. Science 302:1027–1030

Hulme M, Trilsbach A (1989) The August 1988 storm over Khartoum: its climatology and impact. Weather 44(2):82–90

Hulme M (1991) The 1990 wet season in Sudan. Sudan Stud (10):21–22

Hulme M (2001) Climate perspectives on Sahelian dessiccation: 1973–1998. Glob Environ Change 11(1):19–29. doi:10.1016/S0959-3780(00)00042-X

JISAO (2012) NOAA GHCN link to the university of Washington joint institute for the study of the atmosphere and ocean (JISAO) Standard Precipitation Index Web Site: http://jisao.washington.edu/data/sahel/022208/sahelrainjjaso18982004.ascii. See also: http://jisao.washington.edu/data/nicholson/. Accessed 22 Apr 2012

Kraus EB (1977) Subtropical droughts and cross-equatorial energy transports. Mon Weather Rev 105:1009–1018

Nicholson SE, Grist JP (2001) A simple conceptual model for understanding rainfall variability in the West African Sahel on interannual and interdecadal time scales. Int J Climatol 21:1733–1757. doi:10.1002/joc.648

Nicholson SE (1979) Revised rainfall series for the West African subtropics. Mon Weather Rev 107:620–623

Nicholson SE (1986) The spatial coherence of African rainfall anomalies: interhemispheric teleconnections. J Climate Appl Meteor 25:1365–1381

Nicholson SE (1989) Long term changes in African rainfall. Weather 44:46–56. doi:10.1002/j.1477-8696.1989.tb06977

Nicholson SE (1994) Africa rainfall indices, 1901–1994. http://jisao.washington.edu/data/ncholson/. Accessed 15 Feb 2013

Nicholson SE (2009) A revised picture of the structure of the "monsoon" and land ITCZ over West Africa. Clim Dyn 32(7–8):1155–1171

Sutcliffe JV, Dugdale G, Milford JR (1989) The Sudan floods of 1988. Tech Note, Hydrolog Sci 34(3):355–364

UNEP/DEWA-GRID (2010) The precipitation data used for the figures have been extracted by the author from the UNEP/DEWA and GRID Global Precipitation Database GNV174 for the climate normal period 1951–1980, last updated on June 5, 2010. Original global version available @ http://www.grid.unep.ch/data/download/gnv174.zip. Accessed 28 Jun 2010

Wang G, Eltahir EAB, Foley JA, Pollard D, Levis S (2004) Decadal variability of rainfall in the Sahel: results from the coupled GENESIS-IBIS atmosphere-biosphere model. Clim Dyn 22:625–637. doi:10.1007/s00382-004-0411-3

WMO Normals (2012) ftp://dossier.ogp.noaa.gov/GCOS/WMO-Normals/RA-I/SU/. Accessed 24 Nov 2012

Zhang R, Delworth TL (2006) Impact of Atlantic multidecadal oscillations on India/Sahel rainfall and Atlantic hurricanes. Geophys Res Lett 33:L17712. Doi: 10.1029/2006GL026267

Chapter 3
Interannual and Interseasonal Variations in Monthly Rainfall

Abstract Rainfall in the East Sahel is characterized by interannual variabiity and by interseasonal variability, both of which overshadow effects of long-term trends. Using monthly values for more than 100 years of record for three standard gauge sites in the East Sahel—Gedaref (GED), Kassala (KAS), and Khartoum (KHA)—having substantially different annual totals, selected metrics for characterizing the interseasonal variability of precipitation are described and compared. Annual totals *decrease* from south ($P_{annual}^{GED} = 600$ mm) to north ($P_{annual}^{KHA} = 130$ mm), as does the seasonal duration of the monsoon, whereas the relative interannual variability of monthly precipitation *increases*. Dominant rainfall months are typically June (J), July (J), August (A), and September(S) for GED (93 % of annual) and for KAS (96 % of annual). However for KHA, the dominant rainfall months are JAS (96 % of annual). Two common paradigms assumed by the climatology community are assessed—one is that August is the month of maximum rainfall, another is that August is the month most responsible for interannual variability of monsoon rainfall. For the East Sahel, neither paradigm is particularly robust, and both need to be applied with some caution to a particular analysis or region.

3.1 General Patterns

Whereas Chap. 2 described the dataset for characterizing the spatial and temporal variability of annual totals, this chapter describes selected metrics for characterizing the interseasonal variability of rainfall from year-to-year. Certain aspects of the nature of this interseasonal variability for the study area are illustrated using monthly values in Fig. 3.1, showing a 31-year representative section of the much longer (>100 year) time series in Fig. 3.2 from the three reference stations. (For the station locations see Fig. 2.1.)

Only 31 years of data are shown so the viewer can visualize details in the interrelation of seasonal variability and interannual variability. The three time series in Fig. 3.1 show the strong seasonal behavior of monsoonal precipitation in

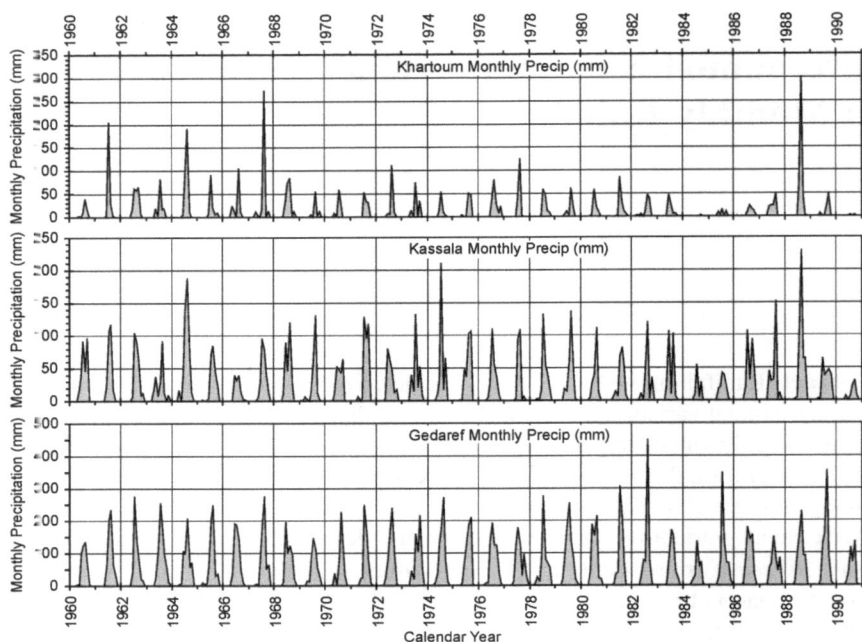

Fig. 3.1 Representative monthly precipitation data from Khartoum, Kassala, and Gedaref for the period of record 1960–1990, coinciding with the 1961–1990 climate normal period

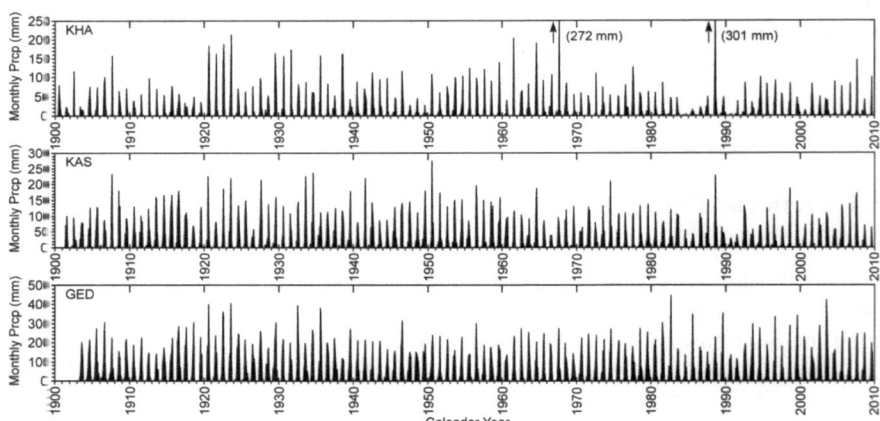

Fig. 3.2 Long-term monthly time series of monthly precipitation values for the three reference stations discussed in the report. Khartoum (KHA) has the widest range of relative values, with two monthly values off-scale: 273 mm for August 1967, and 301 mm for August 1988

this region—there is virtually no rainfall from late fall to mid spring. However, as discussed further below, precipitation can be episodic over the course of the unimodal monsoon season. In addition, there are significant year-to-year or interannual fluctuations in annual precipitation totals. It is noteworthy that whereas the median annual rainfall for Khartoum (130 mm) is significantly less than that for either Kassala (280 mm) or Gedaref (600 mm), there are occasions when the annual monthly maxima for Khartoum significantly exceed those at Kassala or Gedaref, or both (for example in 1961, 1967, or 1988).

3.2 Longer Term Patterns of Interannual Variations in Monthly Precipitation

To show the longer term behavior of interannual variability, Fig. 3.2 has monthly values for the entire 100+ years period of record for each station. For purposes of consistency among the plots, the 1899 data for KHA is not shown. The scale for Khartoum (KHA) is adjusted to maintain a reasonable visual dynamic range, so that two monthly "events" are allowed to go off-scale, but as labeled on the figure, the monthly total for August 1967 is 273 mm, and for August 1988, the monthly total is 301 mm. Both events are associated with intense flooding in the vicinity of Khartoum.

Before the year-by-year details of the respective time series in Fig. 3.2 are discussed, one will note the patterns of multi-decadal modulations of monthly totals, some of which are identifiable in the annual data mentioned in Chap. 2. A number of these long-term modulations are documented in the literature and explained by the teleconnections between local climate and long-term fluctuations of global sea surface temperature (SST) adjacent to the continent in the Atlantic, the Gulf of Guinea, and the Indian Ocean, or from El Nino/La Nina events in the Pacific half a world away (Nicholson 2011). However, the type of local differences apparent in the time series of Fig. 3.2 (as well as in the annual time series in Chap. 2) are more difficult to understand in terms of these teleconnections, and pose significant challenges to climate modelers and forecasters. The pattern of multi-decadal fluctuations at Gedaref during the early part of the twentieth century is particularly striking, where the envelope of the annual monthly maximums suggest a periodicity of 13–15 years. At the same time, it is noteworthy that these long-term fluctuations are asynchronous—or, at least, out-of-phase—across the three-station array. Clearly other, more localized factors are tending to disrupt the coherence of the long-term signal across the three-station array.

On a shorter time scale, the spatial variation of the intensity of floods and droughts is evident in Fig. 3.2. Close inspection of the figure shows that the intense August 1967 precipitation event at KHA (272 mm) is associated with only typical rainfall at Kassala for the same month, and only slightly higher-than-normal rainfall at Gedaref. Similarly, the intense monthly event for August 1988

(301 mm) at the northernmost station, Khartoum, is associated with only modest rainfall at Gedaref, the southernmost station, where the most intense monthly rainfall event over a decadal period (1986 to 1995) occurs, *not* in 1988, but the *following year*, 1989. The rainfall for August 1988 at Kassala is higher than normal, but is not the event-of-record, as is the case for Khartoum (Sutcliffe et al. 1989; Hulme and Trilsbach 1989).

This asynchronous behavior of higher-than-normal monthly rainfall events has its counterpart in the asynchronous behavior of lower-than-normal rainfall months as well, resulting in seasonal dry spells and droughts. The year-long 1984 drought, of course, was experienced throughout the Sahel (cf. Chap. 2). However, in 1985, while Khartoum was still experiencing significantly reduced rainfall (as was Kassala), Gedaref had sprung back to above-normal rainfall. Major droughts were experienced by Khartoum in 1939 and 1949. These, however, are quite subdued, or totally absent at Kassala and Gedaref. Such spatial detail is lost when data from many stations over large areas of the Sahel are aggregated—as discussed in Chap. 2—into regional normalized precipitation indices as in the case of the SPI time series (Fig. 2.7), or percent departures from the annual mean (Fig. 2.11).

3.3 Statistics of Seasonal Behaviors of Monthly Precipitation

3.3.1 Results of this Study for the East Sahel

Figure 3.3 uses 100+ years of monthly data to classify and compare the long-term seasonal or intra-annual behavior of rainfall intensities (mm/month) at the three reference stations in the study. The monthly classification plots are constructed by binning monthly totals from all months having data over the respective period of record indicated in the figure. Months with missing data are disregarded. It should be clear from intercomparing the three plots in Fig. 3.3 that the temporal distribution becomes broader (that is the monsoon season becomes longer) as one moves from north (KHA) to south (GED), and monthly total rainfall increases at the respective site. This, of course, is consistent with the annual south-to-north migration of the Intertropical Convergence Zone (ITCZ) and the tropical rainbelt (cf. Chap. 1).

The statistics for the monthly values shown in Fig. 3.3 are numerically summarized by station with related information in Tables 3.1, 3.2 and 3.3. The maximum monthly extreme for all 100+ years of record for KHA is 301 mm, for KAS is 275 mm, and for GED is 449 mm.

The larger relative interannual variability of monthly rainfall for Khartoum (Fig. 3.3, top panel) is quite evident from the wider interquartile (IQ75–IQ25) brackets (relative to the median monthly values for the station) on its monthly distribution of precipitation.

Fig. 3.3 Classification plots
of seasonal precipitation
using multi-year monthly
totals. *Grey-shaded* area is for
the monthly median. The
quartile values, Q75 and Q25,
for the month are also shown

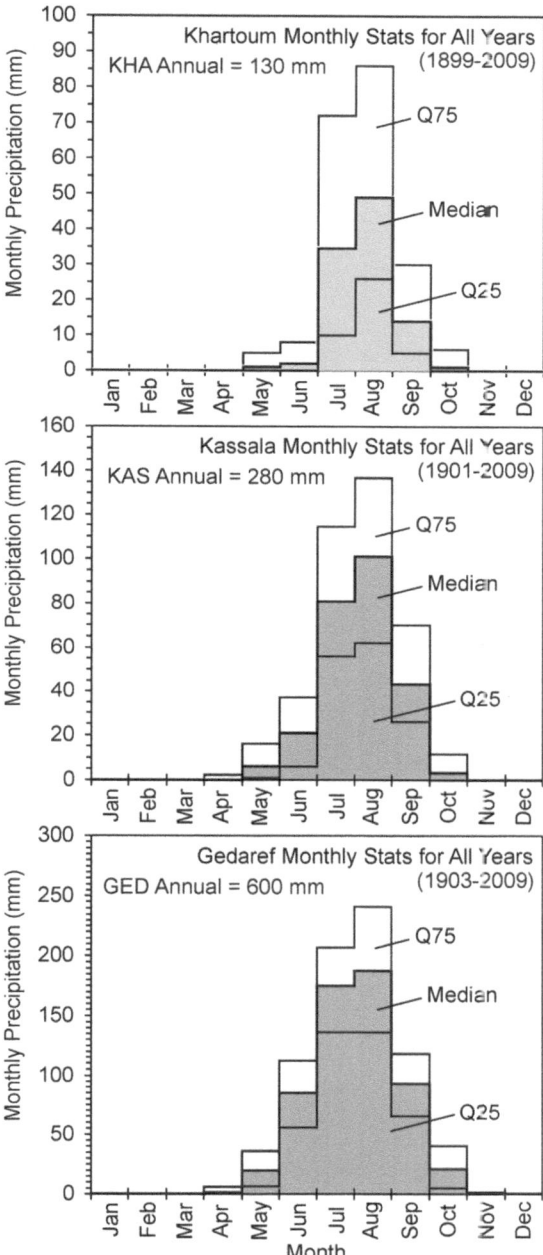

Table 3.1 Gedaref: summary statistics of monthly values

Month	Q25 (mm)	Median (mm)	Q75 (mm)	IQ disper.	% annual
JAN	0	0	0	–	0
FEB	0	0	0	–	0
MAR	0	0	0	–	0
APR	0	1	6	–	0.2
MAY	7	20	36	1.42	3.5
JUN	56	85	112	0.65	14.6
JUL	136	175	207	0.40	30.0
AUG	136	188	241	0.56	32.2
SEP	65	93	118	0.57	16.0
OCT	5	21	40	1.67	3.6
NOV	0	0	1	–	0
DEC	0	0	0	–	0

Table 3.2 Kassala: summary statistics of monthly values

Month	Q25 (mm)	Median (mm)	Q75 (mm)	IQ disper.	% annual
JAN	0	0	0	–	0
FEB	0	0	0	–	0
MAR	0	0	0	–	0
APR	0	0	2	–	0
MAY	1	6	16	2.47	2.4
JUN	6	21	37	1.46	8.2
JUL	56	81	115	0.73	31.8
AUG	62	101	137	0.74	39.6
SEP	26	43	70	1.02	16.9
OCT	0	3	12	3.83	1.2
NOV	0	0	0	–	0
DEC	0	0	0	–	0

Table 3.3 Khartoum: summary statistics of monthly values

Month	Q25 (mm)	Median (mm)	Q75 (mm)	IQ disper.	% annual
JAN	0	0	0	–	0
FEB	0	0	0	–	0
MAR	0	0	0	–	0
APR	0	0	0	–	0
MAY	0	1	5	5.25	1.0
JUN	0	2	8	4.00	2.0
JUL	10	34.5	72	1.79	34.0
AUG	26	49	86	1.22	48.3
SEP	5	14	30	1.75	13.8
OCT	0	1	6	6.00	1.0
NOV	0	0	0	–	0
DEC	0	0	0	–	0

The interseasonal variations of precipitation for both Khartoum and Kassala are more asymmetric than for Gedaref, while all three stations have a single annual monsoon cycle, with maximum precipitation in July and August. August is likely to have the greatest precipitation totals at all three sites, although totals for each locale will be dramatically different.

Table 3.1 shows that 93 % of the annual precipitation for Gedaref occurs from June through September (JJAS), with the interquartile dispersion (IQD) varying inversely with the amount of monthly precipitation—months of higher precipitation totals tend to have lower interquartile dispersion. Recall from Chap. 2 that the IQD = (IQ75–IQ25)/Median, so that for the same spread of variability, as the median becomes small in the off-monsoon seasons, the IQD naturally becomes singular. Because of the sudden jump in the IQD for May and October, one should be cautious if needing to include data from these months into the basis for phenomenological, stochastic, or numerical modeling of the seasonal monsoon in this area

Table 3.2 shows that 96 % of the annual precipitation for Kassala occurs from June through September (JJAS). Note the large IQDs prior to June and after September due to the reduced, but sporadic nature of rainfall outside the principal JJAS monsoon season for Kassala.

Table 3.3 shows that 96 % of the annual precipitation for Khartoum occurs in the *3* months from July through September (JAS), with the IQD erupting to enormous values in the months before and after—again, an artifact of small values in the denominator. Thus the statistical distribution of the principal months of annual rainfall for Khartoum is of shorter duration than for either Kassala or Gedaref. Moreover, it is readily apparent that the interquartile dispersion for all months at Khartoum—including the principal monsoon months: JAS—is significantly greater than for the other stations.

3.3.2 Comparing Local Seasonal Patterns with Prior Work

The three-station study area in this report is in one of the sub-regions in East Africa identified by Nicholson and Selato (2000) as EA1 (or East Africa 1), based on what they feel is a similarity of rainfall styles among a cluster of stations. The seasonal data in Fig. 3.3 clearly have the unimodal annual precipitation pattern for all three stations that Nicholson and Selato report for EA1.

In Fig. 3.4, the three-station monthly percent of expected annual precipitation values have been combined into a single composite using a procedure similar to that used by Nicholson and Selato. The local three-station composite is then overlaid onto Nicholson and Selato's regional EA1 composite. As shown in Fig. 3.4, the seasonal distribution of precipitation for the local three-station study area in the East Sahel is significantly more compact seasonally (largely the 4 months: JJAS) than the regional synthesis of Nicholson and Selato.

This is because some of the stations used by Nicholson and Selato were further to the south than the current study area, and recorded a contribution from lower

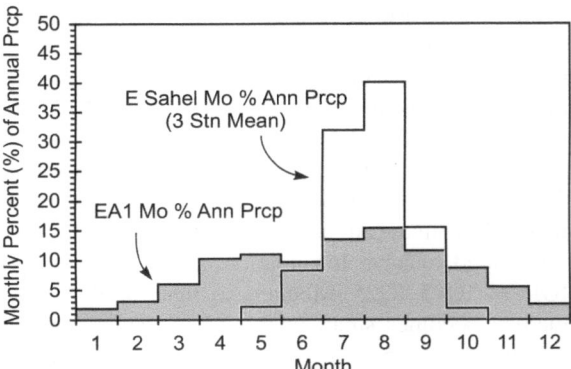

Fig. 3.4 Comparing the seasonal distribution of the monthly percent of annual precipitation for our study area (three-station mean) with the regional analysis of EA1 of Nicholson and Selato (2000)

latitude—more humid stations having a broader distribution of seasonal rainfall. While the result is not surprising, it underscores the role that spatial variability plays in the temporal variability of the delivery of water to the East Sahel, and the necessity of adjusting one's database to the precise geographic region of interest.

3.4 Statistics for the Month of Annual Maximum Monthly Totals

An important metric for monitoring the annual to-and-fro meridional march of the ITCZ—and the position of its associated tropical rainbelt—is the timing of the maximum monthly intensity of precipitation during the year. A common assertion among climate workers is that August is typically the "wettest" month, and has the highest variability, a paradigm that has proven beneficial to the analysis of monsoon rains with respect to the ITCZ in the West Sahel (Dennett et al. 1985; Nicholson and Palao 1993; Nicholson 2009). In this and the following section, this concept is considered in some detail. Consider first the assertion that August is the wettest month.

Clearly this notion concurs with the statistics in Fig. 3.3 for the selected stations in the East Sahel. But a tendency for the maximum to occur in August does not require that the month of August is *always* the month of highest monthly rainfall for any specific year. The latter *could* be required if rainfall were *synchronous* throughout the season such that during a wetter year all months had a proportionately higher rainfall, and during a drier year all months had a proportionately lower rainfall. But August being the wettest month is not required if precipitation is *asynchronous* across the season, and the timing of the annual *seasonal* maximum has an interannual variability—sometimes the month of maximum rainfall is August, sometimes July, etc.

The possibility of such asynchronous behavior is apparent in the monthly time series of Figs. 3.1 and 3.2. Although it may be difficult to resolve the month-by-month values in Fig. 3.1, the 2 years, 1986 and 1987, for the Kassala monthly data illustrate this point quite well.

The following list shows the three (JAS) monthly values for Kassala for two sequential years:

$$1986 : \text{Jul} = 108 \text{ mm} \quad \text{Aug} = 31 \text{ mm} \quad \text{Sep} = 94 \text{ mm}$$
$$1987 : \text{Jul} = 31 \text{ mm} \quad \text{Aug} = 152 \text{ mm} \quad \text{Sep} = 2 \text{ mm}$$

In 1986, the maximum monthly rainfall occurs in July. Not only is the rainfall significantly reduced in August, the August value is only 30 % that in September, and both July and September are significantly higher than August. In 1987, the monthly rainfall is more typical of the rainfall patterns expected from Fig. 3.3 and Table 3.2.

The potential for asynchronous rainfall throughout the season is also evident from inspecting the monthly quantile distributions in Fig. 3.3. For all three stations, the Q75 quantile for July is larger than the median value for August, and the Q25 quantile for August is less than the median value for July. This is consistent with (it allows, although does not require) a year-to-year variability in the timing of the peak seasonal rainfall.

To assess the frequency with which August is the dominant month for precipitation, monthly totals were selected from all years of record having 100 % coverage during the 100+ year period of record (in other words, no months were "filled" with proxies for missing data). The month of maximum rainfall was identified for each year, then the month of each annual maximum was binned by month of the year, with the results plotted in Fig. 3.5, and summarized in Table 3.4.

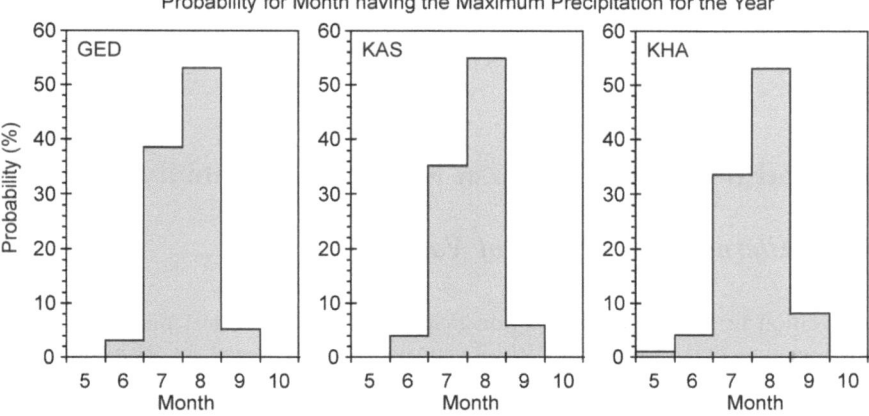

Probability for Month having the Maximum Precipitation for the Year

Fig. 3.5 Probability that a month will have the maximum monthly precipitation for the year

Table 3.4 Probability for a month to have the annual maximum monthly precipitation

Month	Gedaref (%)	Kassala (%)	Khartoum (%)
5	0	0	1
6	3	4	4
7	39	35	34
8	53	55	53
9	5	6	8
July August ratio	74	64	64

For GED, there was a total of 1,152 monthly values, of which 576 fell within the extended monsoon season (MJJASO). For KAS, there was a total of 1,224 monthly values, of which 612 fell within the monsoon season (MJJASO). For KHA, there was a total of 1,176 monthly values, of which 588 fell within the monsoon season (MJJASO). For each year, there is, of course, only one maximum monthly value, so the population of samples is ideally drawn from 108 samples (or years of record) for GED, 110 samples for KAS, and 112 samples for KHA (see Chap. 1, Table 1.1).

What may be surprising in Fig. 3.5 and Table 3.4 is the similarity of the statistics—the shape of the curves—among the three stations. Discounting the probabilities of 3 to 4 % for the maximum monthlies to occur in June, and the 5 to 8 % for the maximum monthlies to occur in September, the noteworthy feature of the analysis is the concordance among the three stations for the results from July and August. For July, the frequency of occurrence for GED, KAS, and KHA of maximum monthlies is 39, 35, and 34 %, respectively. For August, the frequency of occurrence for GED, KAS, and KHA of maximum monthlies is 53, 55, and 53 %, respectively. The ratios of the probabilities of maximum occurrence in July to those in August are given by station in the last line of Table 3.4. The combined probability that the annual monthly maximum will occur in either August or July is GED: 92 %; KAS: 90 %; KHA: 87 %. The evidence in Table 3.4 makes the case that the expected timing of the annual peak of the monsoon season in the East Sahel seems to be a 2-month window, rather than the 1-month window used for some studies in other areas of the Sahel.

3.5 Statistics for the Month of Maximum Variability

3.5.1 Alternative Measures of Variability

In addition to the common assertion that August is the wettest month, there are often claims that the magnitude of year-to-year differences in August rainfall contributes disproportionately to the interannual variability of precipitation for the entire year. In fact, there is a tendency for some Sahelian studies to consider only the interannual variability of precipitation in August at the exclusion of other times

Table 3.5 Variability of monthly and annual JAS totals for GED, KAS, and KHA

1	GED				KAS				KHA			
2	GED (JAS) annual total: 456 mm				KAS (JAS) annual total: 225 mm				KHA (JAS) annual total: 98 mm			
3 Month	% annual	IQD	IQR (mm)	RV	% annual	IQD	IQR (mm)	RV	% annual	IQD	IQR (mm)	RV
4 JUL	30	0.40	71	0.26	32	0.73	59	0.32	34	1.79	62	0.48
5 AUG	32	0.56	105	0.59	40	0.74	75	0.51	48	1.22	60	0.45
6 SEP	16	0.57	53	0.15	17	1.02	44	0.17	14	1.75	25	0.08
7 Seasonal (JAS)	78	0.71	103	1.00	88	0.99	70	1.00	96	1.73	52	1.00
8 Interannual (JAS)		0.33	151			0.50	119			0.80	94	
1	2	3	4	5	6	7	8	9	10	11	12	13

of the season. The analysis in this section, as summarized in Table 3.5, suggests that the interseasonal, as well as the spatial variability may be more complicated, at least for the East Sahel. The table contains several metrics for characterizing such variabilities for the three stations, GED, KAS, and KHA, for the 3 months: July, August, and September (JAS).

Row 1 in Table 3.5 identifies the respective station of record, and Row 2 follows with the total of the monthly medians for the 3-month season—JAS—at each station.

With reference to the respective month in Column 1, the variability for each of the months of the rainfall season is summarized in Rows 4, 5, and 6. Properly speaking, this report suggests elsewhere that there are four principal rainfall months (JJAS) for Gedaref and Kassala, and three principal months (JAS) for Khartoum. However, for the parameters developed in this section, and in particular for Table 3.5, a common seasonal time base of JAS provides a more useful comparison of the three stations.

Both the first (% Annual) and second (IQD) data entries for each month and for each station in Rows 4, 5, and 6 are carried over from the entries in Tables 3.1, 3.2 and 3.3 from the interseasonal analysis of monthly totals. The first data entry for the month (% Annual) in Table 3.5 is the percent of annual rainfall expected at the site for the respective month; and is computed as the median of all values for the respective month divided by the sum of the 12 monthly medians. The second data entry for each station is the respective interquartile dispersion (IQD) value for the month. The third data entry for the month at each station is the interquartile range (IQR) defined earlier in this report by the expression IQR = Q75–Q25. The fourth entry for each station—the relative variance (RV)—will be explained shortly, after the other entries are described.

The second row of data from the bottom (Row 7) is the Seasonal (JAS) for the respective composite metric for the 3-month season (JAS) at each station. In Row 7, Columns 2, 6, and 10—the values of the % Annual for the Seasonal (JAS) for the respective stations—is a simple sum of the percentages for the 3 months in the rows above in the respective column. For example, for Gedaref, the months JAS contribute 78 % of the annual rainfall; whereby the same months for Kassala contribute 88 %; and for Khartoum, the 3 months contribute 96 % of the annual rainfall.

The Seasonal (JAS) IQD in Columns 3, 7, and 11, for the respective station, is collectively computed from the Q75, Q25, and the median of the collective ensemble of all available monthly values for July, August, and September. All the values for these 3 months for a particular station are lumped into a single ensemble. For example, for Gedaref with a 107-year period of record, there will ideally be 3 months × 107 years of monthly values, for an ensemble of 321 monthly samples to draw from. With missing data, however, there are only 316 monthly values available for GED. From the 316 monthly values for GED, the median is computed, and the IQR and the IQD.

Row 8, the final bottom row of data in Table 3.5, contains a set of parameters based on first reducing the monthly JAS values for each year of record to an annual total of the 3 months for the year. Thus for each station, there is only one JAS total per year. For GED, for example, 102 annual samples will be used, since five of the years in the 107-year long period of record had one or more missing months during the JAS season. After the JAS monthly values for each year are collected into a time series of annual totals for the respective station, the IQR (mm) of annual rainfall is computed from 100+ annual samples for the respective station, and provided in Columns 4, 8, and 12. The IQR divided by the median of the 3-month total, provides the IQD values given in Columns 3, 7, and 11 in Table 3.5. The IQR is a measure of the actual or absolute interannual variability (in mm) of the total rainfall in the three rainfall months, JAS, whereas the IQD is a measure of the relative variability with respect to the expected JAS season total for the respective station. It is noteworthy that the interquartile range (IQR) varies in *direct proportion* to the total monthly rainfall (Rows 4, 5, and 6), and in proportion to the seasonal (JAS) total rainfall (Row 8). The interquartile dispersion (IQD), however, varies *inversely* with the seasonal total rainfall for all the cases in Table 3.5.

As an aside, recall that the annual totals in Row 2 are the sum of the three (JAS) monthly medians for each station, such that for GED: 456 mm; KAS: 225 mm; and KHA: 98 mm. Each of the latter would be equivalent to the area under the respective curve in Fig. 3.3 for the 3 months (JAS). By comparison, the *median* (considered here to be the most robust estimator of the expectation) of the 100+ annual totals of the three (JAS) monthly values is for GED: 461 mm; KAS: 240 mm; and KHA: 118 mm. The *arithmetic mean* of the 100+ annual totals of the three (JAS) monthly values is for GED: 471 mm; KAS: 242 mm; and KHA: 133 mm. The message is that one should be clear as to what metric they are employing in a particular application.

3.5.2 Relative Variance

Returning to the fourth monthly entry for each station, RV is the "Relative Variance," a parameter used here to define the effect that the variability of rainfall for a given month has on the variability of the rainfall over the entire season for the station. The parameter follows from a basic proposition from the theory of random variables that the variance of a set of variates, where each variate is composed from a set of subsamples, is the sum of the variances of each of the respective subsamples. In other words, the variance of annual rainfall is the sum total of the variance of the three monthly variances. The RV is defined in this report in such a way so as to relate the variance of each of the three (JAS) monthly subsets to the total variance of annual values (the sum of the three monthly variances). For a strictly normal distribution, the variance is the square of the standard deviation (sd^2), so that, for example, the relative variance (RV) for August, using the variance of the 3-month season—JAS—might be defined by

$$RV_{Aug} = \frac{\left(sd_{Aug}\right)^2}{\left(sd_{Jul}\right)^2 + \left(sd_{Aug}\right)^2 + \left(sd_{Sep}\right)^2} \tag{3.1}$$

However, for data with outliers, a more robust estimator of the range of expected values is the interquartile range (IQR = Q75–Q25). Recalling from elementary statistics that for a normal distribution, 50 % of the distribution falls within ±0.6745 sd of the mean, we have the following numerical relation

$$2 \times (0.6745\,sd) = Q75 - Q25 = IQR \tag{3.2}$$

Hence, the interquartile range is related to the standard deviation of a normal (or Gaussian) distribution by a numerical constant ($sd = 0.7413 \times IQR$), so that, within this numerical constant, the square of the IQR can be used as a proxy for the Gaussian variance, which is, in turn, the square of the standard deviation, $\left(sd_\mu\right)^2$, where μ denotes the μth month. The parameter $\left(IQR_\mu\right)^2$ will analogously be referred to as the "interquartile variance" for the μth month (which is nothing more than the square of the interquartile range referenced to a specific month). Thus, as a robustified proxy for (3.1), the relative variance (RV) will be defined for each month as the ratio of the square of the interquartile range for the respective month (over the 100+ year period of record), normalized by the sum of the squares of the interquartile range for the 3 months of the rainy season (JAS). As an example, for August,

$$RV_{Aug} = \frac{\left(IQR_{Aug}\right)^2}{\left(IQR_{Jul}\right)^2 + \left(IQR_{Aug}\right)^2 + \left(IQR_{Sep}\right)^2} \tag{3.3}$$

If other combinations of months were used, a general expression for the RV for the respective month would be

$$RV_\mu = \frac{(IQR_\mu)^2}{\sum_{\mu=1}^m (IQR_\mu)^2} \qquad (3.4)$$

where m is the total number of months, and μ is an integer index identifying the month (here, $\mu = 1$, for June; $\mu = 2$, for July; etc.). A useful attribute of the RV is that

$$\sum_{\mu=1}^m RV_\mu = 1.0 \qquad (3.5)$$

regardless of the station, and its associated expected total rainfall. To illustrate computing the parameter, in Table 3.5 for GED, the IQR in the table is 71 mm for July, 105 mm for August, and 53 mm for September. Thus, the relative variance for August, RV_{Aug}, is

$$RV_{Aug} = \frac{(105)^2}{(71)^2+(105)^2+(53)^2} \approx 0.59 \qquad (3.6)$$

as entered into Table 3.5 in Row 5, Column 5. The sum of the RVs for a site is unity, as shown in Row 7, so that the monthly RV is a useful parameter to describe the expected partitioning of the variability of rainfall for a specific month relative to the contribution of all of the months contributing to the variability of the respective annual totals for a specific station. The square root of the ratio of a pair of RVs is a proxy for the ratio of the standard deviations of the respective monthly totals.

Inspecting the relative variances (RVs) for Gedaref in Table 3.5, it is apparent that approximately 59 % of the interannual variance of annual totals is associated with August, which might justify declaring August, at this site, as being responsible for "most" of the interannual variability. However, 41 % of the interannual variance is contributed by July and September. Another way of comparing August to the other 2 months is to compute the square root of the ratio of 0.41 to 0.59. Doing so suggests that the interannual variability of rainfall in July and September supplies approximately 83 % as much as August to the interannual variability of annual totals as represented by the IQD of annual totals.

Inspecting the results for stations having lower annual precipitation—KAS and KHA—indicates a shift in the balance of variabilities. At Kassala, the ratio of variances for July to August is 0.32 to 0.51, so that July is supplying 63 %—more than half—of the variance of August. Moving on to Khartoum, the ratio of variances for July to August is 0.48 to 0.45, suggesting that variability of the monsoons in July in this section of the East Sahel play at least as large a role in the annual variability of rainfall as do the monsoons in August.

Thus, any general assertion that August is the month responsible for the greatest variability of annual rainfall in the Sahel—at least the East Sahel—needs to be regarded with some caution, and assessed in terms of its implications for a particular analysis or application, and depending on the region.

References

Dennett MD, Elston J, Rodgers JA (1985) A reappraisal of rainfall trends in the sahel. J Climatol 5:353–361. doi:10.1002/joc.3370050402

Hulme M, Trilsbach A (1989) The August 1988 storm over Khartoum: its climatology and impact. Weather 44(2):82–90

Nicholson SE (2009) A revised picture of the structure of the "monsoon" and land ITCZ over West Africa. Clim Dyn 32(7–8):1155–1171

Nicholson SE (2011) Dryland climatology. Cambridge University Press, Cambridge

Nicholson SE, Palao IM (1993) A re-evaluation of rainfall variability in the Sahel. Part I. Characteristics of rainfall fluctuations. Int J Clim 13:371–389

Nicholson SE, Selato JC (2000) The influence of La Nina on African rainfall. In: J Clim 20(14):1761–1776

Sutcliffe JV, Dugdale G, Milford JR (1989) The Sudan floods of 1988, Tech. Note. Hydrol Sci 34(3):355–364

Chapter 4
Intra-Seasonal Patterns of Rainfall from Daily Values

Abstract To understand rainfall climatology at the short period end of the spectrum, it is necessary to understand patterns in the day-by-day temporal and spatial variability of rainfall. Rain-fed agriculture thrives best under low-intensity, high-frequency rain days. Plans for flood management and erosion control need to account for the likelihood of intense, though infrequent, storm days. Unfortunately, the years for which daily data are available are limited for the three study sites in the East Sahel (Gedaref, Kassala, and Khartoum), and gauge data for this area at intervals of less than a day are non-existent in the WMO-GHCN archives. Daily gauge data are largely available before 1991 and after 1957. Recognizing their limited period of record, this chapter draws statistics from 6,570 daily values for Gedaref (GED); 12,045 daily values for Kassala (KAS); and 10,950 daily values for Khartoum (KHA). Several common procedures are described for analyzing the frequency and expected intensity of rain days—methods that should have application to other databases as they become available. As determined from prior studies by others in the Sahel (primarily the West Sahel), the three-station daily data in the East Sahel show that while most of the rain days have low intensity (mm/d), most of the annual rainfall is contributed by the fewer, high intensity events likely associated with local scale intense convective storm systems. The monsoon season in the Northeast Sahel (KHA) tends to be 20–30 days shorter than the season to the south at KAS and GED. The mid-seasonal behaviors of the three datasets are quite similar, with close to identical values for the cumulative percent (60 %) of annual precipitation by Day 225 (Aug. 13). Analytical "models" are developed to represent the expected seasonal distribution of daily rainfall using iterative least-squares. The expected timing of peak daily precipitation in percent of the annual total typically occurs about Day 223 (Aug. 11), differing among these three stations by only a matter of a few days.

J. F. Hermance, *Historical Variability of Rainfall in the African East Sahel of Sudan*, SpringerBriefs in Earth Sciences, DOI: 10.1007/978-3-319-00575-1_4, © The Author(s) 2014

4.1 General Patterns of Daily Precipitation

Daily precipitation in the East Sahel is both spatially non-uniform and temporally irregular as illustrated by Fig. 4.1, which is a map of satellite-based rainfall estimates for a representative day in early Aug. 4, 2012. The rainfall estimates in Fig. 4.1 were produced by the U.S. Climate Prediction Center (CPC) using an algorithm that merges rainfall data from three quite different types of sources. One type of data is provided by geosynchronous satellites (e.g. METEOSAT). High-resolution, thermal infra-red imagery is used to identify cloud formations having the coldest temperatures. The coldest clouds are most likely the highest in altitude, and are most likely to produce the greatest rainfall (Arkin and Meisner 1987). The geographic area covered by clouds colder than an assigned threshold is combined with a second type of data: passive microwave backscatter radiometer products from low earth-orbiting satellites. The backscattering information provides estimates of the volumetric ice content (hence, potential rainfall) of clouds. The third

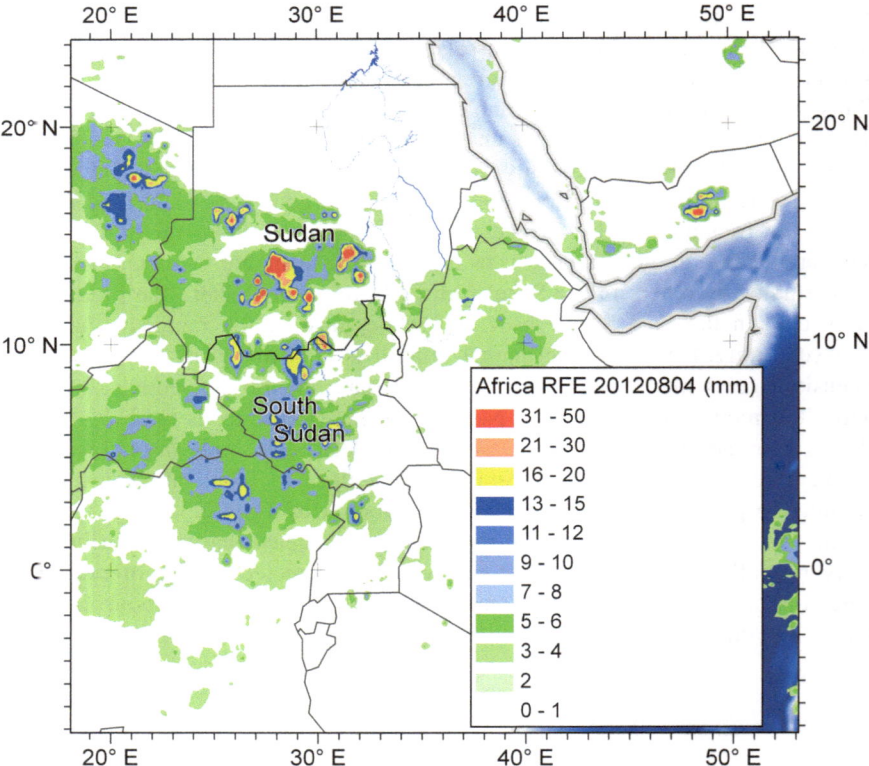

Fig. 4.1 Map of estimated daily (24 h) rainfall for Aug. 4, 2012, based on merged satellite and ground-based data, routinely provided by the Climate Prediction Center (CPC) as an RFE (rainfall estimate) product. See CPC (2012)

type of data is provided by a thousand or more ground-based rain gauges on the African continent that are used to adjust the amplitude and baseline of the final merged rainfall estimate (Xie and Arkin 1996). The CPC RFE 2.0 product for Africa provides daily (24 h) estimates of expected rainfall over a surface grid of 0.1 × 0.1 degree cells (or pixels) spanning the continent from 40°S to 40°N, and from 20°W to 55°E (CPC RFE 2.0 2012). Only a portion of the daily-produced map is shown in Fig. 4.1.

While products like the RFE 2.0 are proving to be a valuable resource for regional scale studies in Africa over time frames of months to years (Gebremichael and Hossain 2010), the utility of satellite rainfall estimates for monitoring single storm events at the scale illustrated in the figure is yet to be determined (Hermance and Sulieman 2013). As mentioned in Chap. 1, one of the most fundamental characteristics of Sahelian rainfall is the extreme local scale and short duration of the type of convective rainfall events characteristic for the region (El Tom 1975; El Gamri et al. 2009). The spatial scale of these events is illustrated in Fig. 4.1. This and the following chapter will show that historical daily gauge data provide significant insight into the spatial and temporal variability of these types of storm events. It should become clear that the need for high-quality ground-based gauge data to address the type of water management issues identified in this report—from rain-fed agriculture to flash flood mitigation—will continue for some years to come.

This chapter will assess interseasonal and interannual patterns of daily rainfall at the three sites in the East Sahel shown in Fig. 4.2. The 200 m elevation contours in the figure show that these three stations are, more or less, at similar elevations on the rolling plains of Sudan, northwest of the Ethiopian and Eritrean Highlands. The most striking difference between them is that the stations lie along a sharp north-by-northwest gradient in annual rainfall of approximately 2 mm/yr/km (see Chap. 2)—which is to say that over a distance of 200 km, along the strike of this gradient, annual totals differ by 400 mm/yr.

The type of temporal variability of rainfall intensities that one should expect among these three stations is shown in Fig. 4.3 for daily rainfall data from a typical year. Global Historical Climate Network (GHCN) data from calendar year 1973 are chosen for the example in the figure since annual totals for that year are relatively characteristic for the respective stations (see Fig. 4.4). The long-term expected annual totals used in this chapter are based on those computed in Chap. 2 from time series of 100 years or longer. Total precip for 1973 at GED is 596 mm (expected value = 600 mm); at KAS, the 1973 rainfall is 260 mm (expected value = 280 mm); and at KHA, the 1973 rainfall is 159 mm (expected value = 130 mm). Figure 4.4 compares 1973 with other years for the available periods of record for daily data used in this report.

It should be readily apparent from Fig. 4.3 that rainfall can be quite episodic and spatially uncorrelated among all three sites, as is also evident in the satellite-based rainfall distribution in Fig. 4.1. Storm events are often interlaced with extended storm-free periods during the monsoon season that typically commences after Day 150 (May 30), and does not extend beyond Day 300 (Oct. 27), which accords with the seasonal analysis of monthly values in Chap. 3.

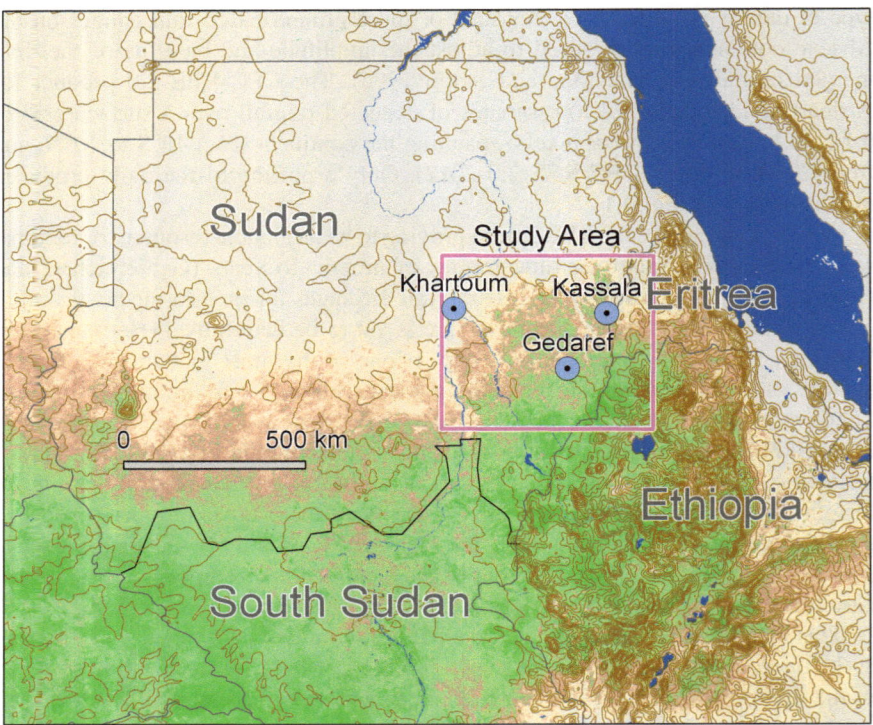

Fig. 4.2 Locations of the three WMO rainfall gauges used in this study. The outlined study area
is the same as in Figs. 1.1 and 1.2. Contours are elevation at 200 m intervals. The background
vegetation map is from a USGS NDVI database for September 1992 (see Fig. 1.2)

4.2 Daily Precipitation Records of Opportunity

There is a fundamental problem with acquiring adequate gauge data in the East
Sahel largely due to a combination of factors, including conflicts, political strife,
and local economics, among others. GCOS (2004) observes that the flow of ter-
restrial rainfall data to the international data centres and the user community, from
a number of members of the terrestrial-observing networks, is not adequate, and in
fact is often quite poor. In the case of rainfall data from the GHCN, many years are
simply missing. This is particularly true of daily precipitation data from Sudan
after 1990, when data coverage dropped precipitously. Even before that time,
however, the number of active stations providing historic data varied from decade
to decade. Therefore, the daily data available for this report—in fact data available
for the ambitious programs envisaged by GCOS (2004) for the future—have
nowhere near the coverage as the monthly data referred to in earlier chapters. The
data that are available are referred to here as "records of opportunity," and
because of the limited periods of record available, any statistical inference needs to

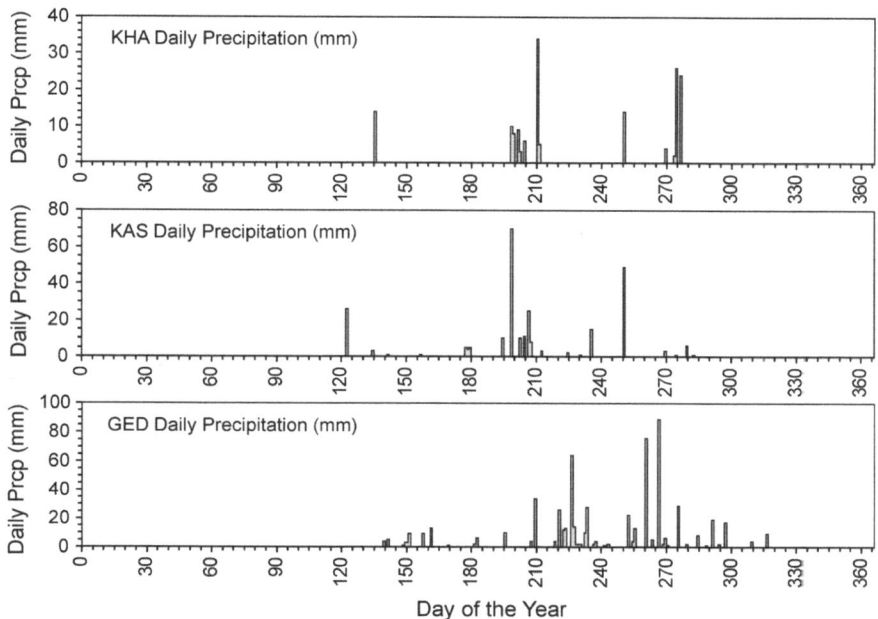

Fig. 4.3 GHCN daily precipitation values for a typical year—1973—recorded at the three reference stations. For WMO station IDs and coordinates, see Table 1.1. Data from GHCN-Daily (2012)

be drawn with some caution. Figure 4.4 shows the daily datasets from the three reference stations to be used in this report for which data are contiguous and of relatively good quality (EarthInfo 2010). Each time series is of different length. For purposes of inter-station comparisons, the vertical scale on the left (daily total) is the same for all plots, and the vertical scale on the right (annual total) is the same for all plots.

The coverage of these stations for the periods of record in each panel of Fig. 4.4 is, from north to south (top-to-bottom in the figure), 99 % for Khartoum, 98 % for Kassala, and 99 % for Gedaref. Each time series is the longest, continuous, overlapping record for these stations that is currently available through the Global Historical Climate Network (GHCN-Daily 2012) of the World Meteorological Organization (WMO). As in the case of the monthly data from Kassala for the period of 1961–1989 in Fig. 1.4, the daily values for Kassala in Fig. 4.4 also show the major storm and consequent flood events of 1964, 1974, and 1988. It is noteworthy, however, that the extreme storm event in 1983 for KAS (Fig. 4.4) did not leave a significant imprint on the monthly values in Fig. 1.4. This has important consequences for local planners as well as climate scientists: The history of short-term flash floods may not be totally recoverable from monthly data alone. Also in Fig. 4.4 is the low rainfall at KAS associated with the regional drought

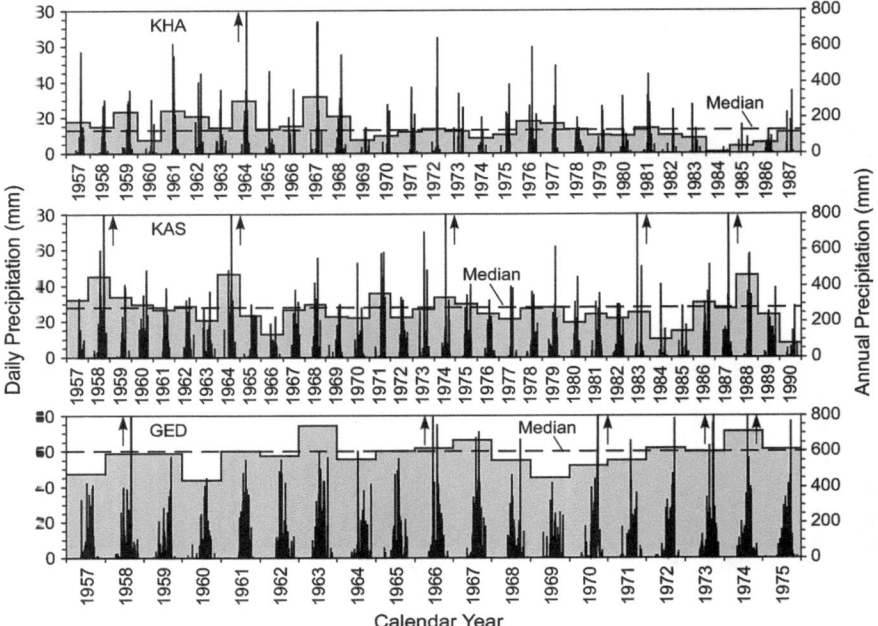

Fig. 4.4 Daily time series employed in the present analysis plotted to the same scale (*left axis*). Dashed lines (referred to the *right axis*) show the long-term median or expected annual values from monthly values (mm) from Chap. 1. *Grey-shaded* step plots (referred to the *right axis*) show the annual monthly totals (mm) from Chap. 2. *Vertical arrows* indicate events where the daily value exceeds 80 mm (the actual values for these extreme events are provided in Table 4.1)

Table 4.1 Daily precipitation intensities greater than 80 mm/d

Date (365 day/ year)	GED Prcp (mm/d)	Date (365 day/ year)	KAS Prcp (mm/d)	Date (365 day/ year)	KHA Prcp (mm/d)
1958 670	172	1958.719	90	1964.618	84
1966 475	89	1964.629	81		
1966 478	89	1974.536	87		
1970 749	100	1983.423	105		
1973 730	89	1987.645	90		
1974 621	81				

events of 1966, 1984, and 1985. These are also quite apparent for the monthly time series from KAS in Fig. 1.4. The correlation, or lack thereof, of other events at these three stations will be considered later in this report.

4.3 An Overview of Daily Precipitation Patterns

Figure 4.5 shows example datasets for a common time base comparing representative daily precipitation totals (mm) at GED, KAS, and KHA along with the long-term median annual precipitation (dashed line) and the annual totals computed from monthly values (the step graph with grey background).

For precise comparison, the data from each station are plotted using a common vertical scale on the left for daily values, and a common vertical scale on the right for annual values and long-term median values. These five years of data are plotted in such detail to emphasize several points reported in the literature from studies elsewhere along the African Sahel. First, rainfall tends to be episodic, rather than typically lasting over prolonged periods. Second, while occasionally some storms appear to be temporally synchronous at two or more stations, other events are quite isolated. The latter type of spatial intermittency is apparently quite typical in the West Sahel and is described in some detail by Balme et al. (2006), who used a substantially greater density of stations. In the East Sahel, much of the rainfall derives from locally convective storms as well (El Toms 1975), so that the lack of spatial coherence of events is not surprising. The few synchronous events in the East Sahel that appear in at least two of the time series in Fig. 4.5 would be

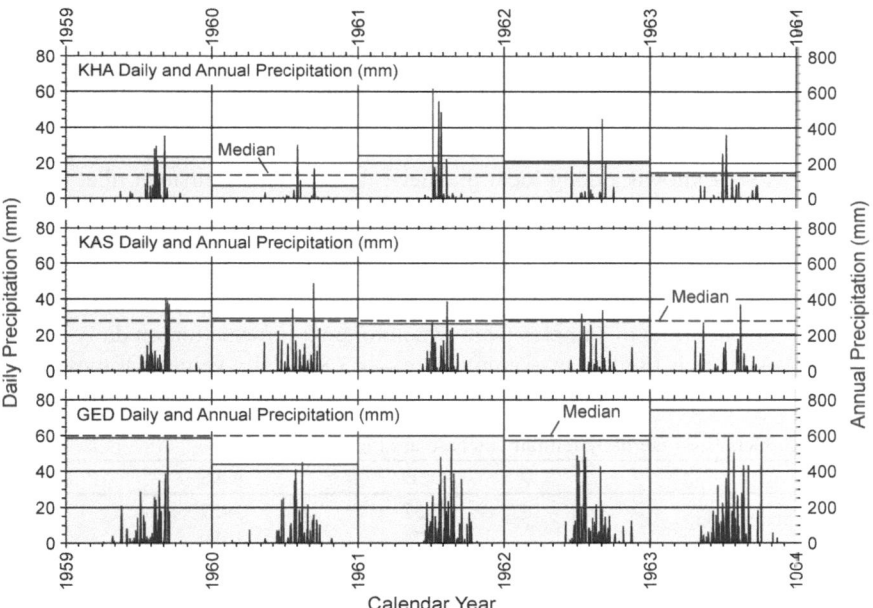

Fig. 4.5 Example dataset comparing representative daily precipitation (*left axis* in mm) among the three stations—GED, KAS, and KHA. Referred to the *right axis* is the long- term median annual rainfall (*dashed line*) and each annual total (*right axis*). Ticks along the time axis are in years and months, Jan.–Dec. Years labeled at their start

expected to be associated with larger-scale mesoscale or frontal-based weather systems. Third, despite the significant and systematic differences in the expected annual precipitation among these stations—600 mm at Gedaref, 280 mm at Kassala, 130 mm at Khartoum—storm day totals (mm) from single events may be as high or higher at any of the stations compared to the others. This accords with the reports of others (Balme et al. 2006; Lebel and Ali 2009) who note that differences in seasonal totals (mm/y) in the West Sahel are dominated by differences in the frequency of storm events, not necessarily by differences in the intensity of discrete events.

4.4 Statistics of Daily Rainfall Intensities

4.4.1 Frequency of "Rain-Days"

For purposes of this report, a "rain day" (or "storm day") is defined as any recorded 24-h period having rainfall in excess of one mm (more precisely, ≥ 1 mm). As mentioned previously in this report, daily rainfall data available through the WMO are limited. Table 4.2 summarizes the nature of the database from the GHCN available through 2010. The beginning and ending years of the period of record for each gauge are given for the respective gauge, followed by the total number of years and then the number of days in the typical monsoon season from April 1 through Oct. 31. Consider the case of Gedaref: multiplying the number of years (18) by the number of days in the monsoon season (214) provides the total number of days for which it is reasonable to calculate statistics (3,852). Since it is well known among local planners, farmers and herdsman, that there is seldom any rain at all outside of the April–October rainfall season, for the purpose of analyzing the frequency and intensity of rain days in this section, other days of the calendar will be excluded. The population of a town, for example, seldom counts the citizens in the cemetery when its computes per capita income.

Thus the ensemble of samples from which to draw information on daily rainfall for GED—which is used as an example here—is 3,852 daily values. Of these, most

Table 4.2 Details on the daily rainfall database used in the analysis

	GED	KAS	KHA
Begin year	1957	1957	1957
End year	1974	1989	1986
# of year	18	33	30
# days in season (Apr–Oct)	214	214	214
Total # days avail. (Apr–Oct)	3,852	7,062	6,420
Total # days >1 mm	888	874	408
% rain days during season	23	12	6
Expected rain days per year	56	30	15

were without rain, and 888 days had rainfalls ≥ 1 mm, so qualify as *rain days* (or *storm days*). Dividing the number of rain days (888) by the total number of days in the rainy season (3,852), gives the probability that a given day in the season will have rain (in the case of GED: 23 %). Multiplying the total number of days in the season (214) by the latter probability (23 %), implies that GED will typically have approximately 56 rain days during the monsoon season. According to the other columns in Table 4.2, the number of rain days for KAS (30) is approximately half of this, which is consistent with the total annual rainfall at KAS (280 mm) being approximately half that at GED (600 mm). Moreover, the number of rain days at KHA (15) is approximately half again, consistent with an expected annual rainfall at KHA of 130 mm.

4.4.2 Statistics of Frequency Versus Intensity

Recalling the gross differences between the annual rainfall at each of the three study sites from the last section (Table 4.2), this section statistically assesses the relation between the frequency and the intensity (as the daily total) of storm days contributing to the annual mix. Strictly speaking, there is a difference between *intensity*, which is the *rate* of rainfall (having units such as mm per unit time) and *daily total*, which is simply the accumulated depth of rainfall in a given day (having units such as mm). This distinction is extremely important in cases where rainfall is recorded and available for analysis at shorter intervals, such as by the hour, or by the minute. The dynamics of how the rainfall in a storm is distributed in time and rate is essential to understanding the hydrological impact of the event on runoff, containment and erosion (Balme et al. 2006; Lebel and Ali 2009). Unfortunately, the only data available for this part of the Sahel are daily values, so that when speaking in terms of daily values (mm), the difference between the terms *daily total* and *intensity* becomes somewhat blurred when the only measure available for intensity is mm per day (mm/d). One might find it helpful for some applications to use metrics for intensity such as *mm per week* or *mm per dekad* (where 1 dekad $=$ 10 days), but that will not be done in this report. Thus, in practical terms for the purposes of this report, daily totals serve as a proxy for daily intensity.

Figure 4.6 is a composite of two types of metrics for statistically classifying daily intensities over an extended period of record. The left axis and its associated curves depict the probability that, if a rain day occurs, its daily total (in mm) will exceed the value of daily rainfall along the abscissa. The right axis and associated curves depict the cumulative fractional percentage of annual rainfall contributed by rain days having daily totals less than the value along the abscissa.

The respective exceedence probabilities (left axis) for the three gauges, to all intents and purposes, track each other quite well. For example, there is a 20 % probability that daily rainfall will be greater than 15–16 mm for KHA and KAS, and greater than approximately 20 mm for GED. The cumulative fraction of total precipitation contributed by rain days having daily totals less than an assigned

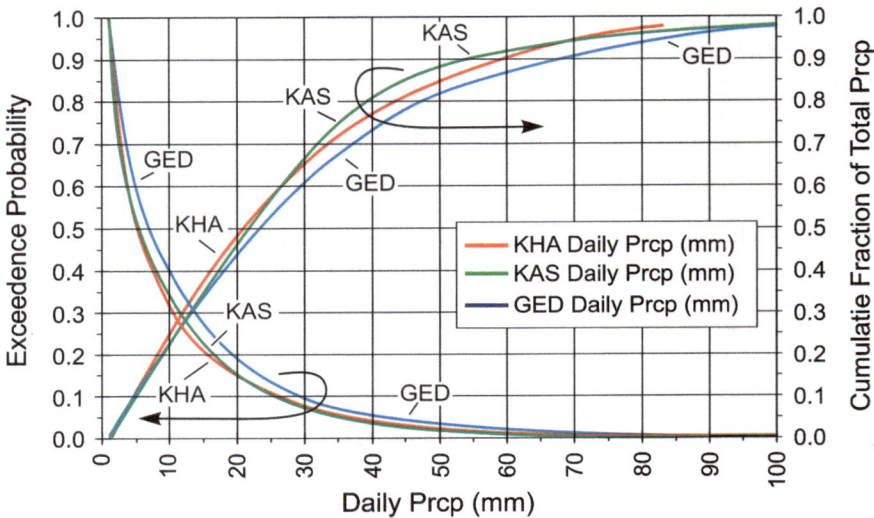

Fig. 4.6 Statistics of daily rainfall totals. The exceedence probability is scaled to the *left*, and the cumulative fraction of the total annual rainfall contributed by events having lesser daily totals is plotted to the *right*. The point to take way from these plots is the relative similarity in the patterns for the respective metric among the three sites

value also track quite well among the three datasets. For example, 40 % of the annual rainfall is contributed by days having intensities less than 15 mm/d for KHA, 17 mm/d for KAS, and 18 mm/d for GED. For purposes of planning and design, considering the limited datasets available for this analysis, these values are virtually the same.

One of the most interesting aspects of these plots is the tradeoff between the frequency of rain days, their intensity, and the contribution of the intensity of a rain day to the annual total rainfall. Consider the section of the graphs where the respective metrics cross each other, nominally in the range along the abscissa of 10–14 mm, or for an intensity of approximately 12 mm/d. The exceedence probability is approximately 30 %, and the cumulative fraction of total rainfall is also approximately 30 %. The implications of this are that 70 % of all rain days have intensities *lessthan* 12 mm/d (or 30 % have intensities greater than 12 mm/d). However, 70 % of the annual rainfall is contributed by rain days having intensities *greater than* 12 mm/d (or only 30 % of the annual rainfall is contributed by rain days having intensities less than 12 mm/d). Simply stated, while most of the rain days have low intensity, most of the annual rainfall is contributed by the fewer, high-intensity storm days. The partitioning of the total precipitation during a monsoon season between frequency of occurrence and rainfall intensity has important implications for a variety of planning exercises. Rain-fed agriculture would ideally thrive under low-intensity, high-frequency conditions; and flood management and erosion control are concerned with the most intense—although infrequent—rain days.

4.5   Extreme Daily Precipitation

In terms of the statistics of extreme intensities in Fig. 4.6, 10 % of the rain days will have precipitation greater than 30 mm/d; 5 % of the rain days have precipitation greater than 40 mm/d. It is assumed here that daily intensities greater than the 95th percentile (40 mm/d) qualify as extreme daily events. The daily maximum—the maximum total rainfall for a single 24-h period during the period of record for which data were available—for GED for this analysis (period of record: 1957–1974) was 172 mm on Sept. 2, 1958. For KAS (period of record: 1957–1989), the daily maximum was 104 mm on June 4, 1983. For KHA (period of record: 1957–1986), the daily maximum was 84 mm on Aug. 14, 1964. (Note that the database used here does not include 1988 during which the storm of record for Khartoum occurred on Aug. 4, with a daily total of 200.5 mm.)

Table 4.3 summarizes the maximum of daily totals by month for the WMO Normal Climate Period of 1961–1990. Values on the order of 70 mm and larger are extremes of record.

4.6   Seasonal (Annual) Patterns of Daily Precipitation

The multiyear time series in Fig. 4.4 have been reordered in Fig. 4.7 to provide a first impression of the expected seasonal distribution of daily precipitation. Following normal convention, any data from February 29th in leap years have been dropped, and a 365-day calendar is adopted for all years. Daily precipitation values have been sorted and overlaid in Fig. 4.7, so that there is a tendency to mask smaller events occurring on the same calendar day for different years. The full period of record in Fig. 4.4 is used for each station in Fig. 4.7, so that the respective number of events for each panel will vary.

An essential point to draw from Fig. 4.7 is the gross inter-event variability of storm totals at each respective station. Moreover, all three datasets show a tendency for a seasonal concentration of precipitation between Day 150 and Day 300, with many events for Khartoum (KHA) having daily totals similar to those at the southern, "wetter" stations.

The bottom panel for Gedaref in Fig. 4.7 shows that for the period of record 1957-1974, the maximum daily rainfall at this station was 172 mm (indicated by the adjacent arrow). According to Fig. 4.4 and Table 4.1, this event occurred in 1958. (The 172 mm maximum in Fig. 4.7 should not be confused with the 100 mm daily maximum in Table 4.3, which occurred in August 1982. The period of record used for Table 4.3 is 1961–1990.)

Fig. 4.7 Overlays of daily precipitation data occurring on the same day over different years for the complete period of record for the respective station (see Fig. 4.4)

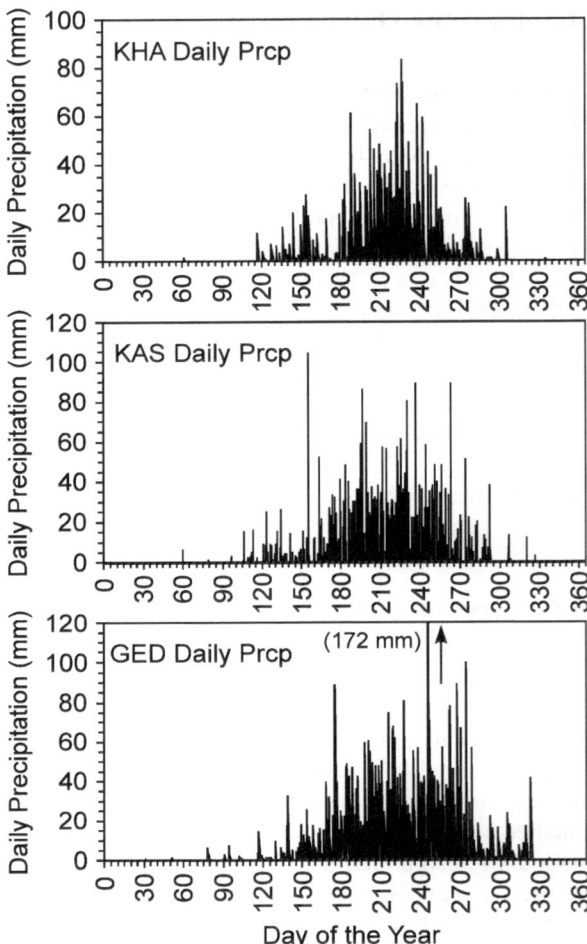

Table 4.3 Maximum daily rainfall (mm) for all days in Climate Normal Period: 1961–1990 (WMO Normals 2012)

Month	GED (62752)		KAS (62730)		KHA (62721)	
	Daily max (mm)	Year of max	Daily max (mm)	Year of max	Daily max (mm)	Year of max
June	89	1980	104	1983	28	1983
July	90	1981	87	1974	62	1961
Aug	100	1982	81	1964	200	1988
Sep	77	1975	59	1971	45	1962
Max	100	Aug	104	June	200	Aug

4.7   Seasonal Accumulation of Annual Precipitation

The seasonal accumulation of annual rainfall in this part of the Sahel is illustrated in Fig. 4.8. The daily precipitation data from each station are ordered by day of the year, and normalized by the rainfall total for all days over the period of record for each respective station. In the following equation,

$$P_m(\% \, annual) = \frac{\sum_{i=1}^{m} \sum_{j=1}^{k_i} P_{ij}}{\sum_{i=1}^{365} \sum_{j=1}^{k_i} P_{ij}} \qquad (4.1)$$

the index i is the ith day of the year (from $i = 1$–365), and k_i is the number of ith days available. For example, for a 25-year period of record, there should be 25 samples of each day of the year, so that ideally, say, for Day 22, $i = 22$ and $k_{22} = 25$, although missing data might cause k_i to be different for different days of the year. P_{ij} is the daily total (mm/d) for the ith calendar day and the jth sample of precipitation on that calendar day over the total period of record. It should be apparent that the denominator of (4.1) is simply the sum of all daily samples over the total period of record for the respective station. The numerator is the sum of all events occurring from Day 1 through Day m for all the years of record. In other words, relation (4.1) indicates the integration (or summation) of the normalized daily values from Day 1 through Day m, to provide $P_m(\% \, annual)$, which is the cumulative fractional percent of annual rainfall occurring on or before the respective day (Day m) of a representative year.

Later in this chapter, the normalized data from these three stations will be compared in some detail. However, for now, it is instructive to highlight a first-order difference in the seasonal pattern at Khartoum—on the relatively dry, northern edge of the Sahel—with the composite behavior of the relatively wetter stations, Gedaref and Kassala, to the south. Due to the sparsity of daily values at Gedaref, and the fact, as discussed later, that the seasonal shapes of the GED and the KAS distributions are materially the same, it is appropriate to combine the

Fig. 4.8 Cumulative percent of annual precipitation plotted by day of the year for Khartoum (KHA), compared to the weighted (Wtd) composite metric for Gedaref (GED) and Kassala (KAS)

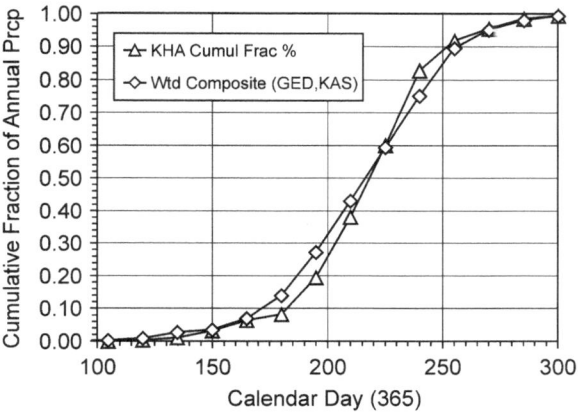

results from Gedaref and Kassala into a single, composite metric. Figure 4.8 shows the daily percentages recomputed as simple cumulative values in bins of 15-day intervals (Day 1–15, Day 16–30, etc.), then plotted at the end of the 15-day steps over the calendar year to Day 365 (leap year values for Feb. 29 are dropped). The composite GED and KAS data were composited by weighting the percent of annual rainfall in each 15-day bin by the respective number of daily samples available, then computing the weighted mean for the respective bin. Data for all 365 calendar days were used in the computation for Fig. 4.8, although to highlight the more significant aspects of the results, only the normalized values from Day 100 through Day 300 are shown.

In Fig. 4.8, seasonal precipitation for the composited data from the higher precipitation, southern stations, GED and KAS, develops earlier in the season, and extends later in the season, compared to the precipitation pattern for the northernmost—and dryer—KHA station, perhaps a maximum difference in timing of up to 10 days in June, with a similar difference in October. The mid-seasonal behaviors of the two datasets are quite similar, with close-to-identical values for the percent of annual precipitation (60 %) on or about Day 225 (Aug. 13). Fifty percent of the annual precipitation is received before Day 216 (Aug. 4) for the GED-KAS composite, and before Day 218 (Aug. 6) for KHA, a difference of two days that, in practical terms, is irresolvable for these data.

4.8 Seasonal Models for Percent of Annual Precipitation Based on Daily Data

For many applications in hydro climatology and water management, it is useful to have a seasonal model to represent the temporal distribution of rainfall throughout the year. Figure 4.9 contains cumulative plots of the fractional percent (or decimal

Fig. 4.9 Fit of Cumulative Weibull functions to the cumulative daily percent of annual precipitation at the three stations

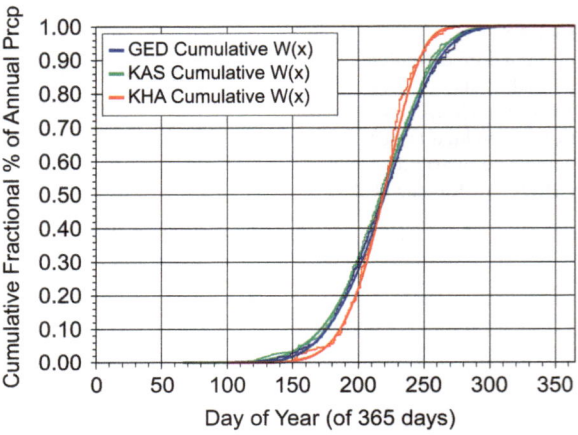

fraction) of rain that has fallen by day of the year for each station as computed from relation (4.1). Each ragged line is the numerical integration (summation) of precipitation at the respective station by day of the year.

The corresponding smooth curve is a least squares fit of a Cumulative Weibull function W(x) to the observed data. This function is widely used in hydrology, and named for Waloddi Weibull, a Swedish Academician who gave the expression wide notoriety (Weibull 1951). Its functional form is defined for this report by

$$W(x, x_o) = 1 - e^{-((x-x_o)/\lambda)^k} \tag{4.2}$$

for the range $x \geq x_o$, whereas for $x < x_o$, $W(x, x_o) = 0$. In the expression (4.2), x_o is an offset or simple translation term, and λ and k are empirical shaping or scaling factors greater than 0.

The data used are taken from the time series shown in Fig. 4.4, after screening for anomalous, apparently misdated, or otherwise suspect daily values. Since the primary application of a seasonal model is to capture the essence of the seasonal pattern, data for the full 365-day calendar year were trimmed to a time period generally consistent with the start-of season (SOS) and end-of-season (EOS) analysis in the previous section, and including the principal rainfall months at each site. The resulting trimmed databases included daily rainfall for the calendar days listed in Table 4.4.

The last entry (the last column) in each line in Table 4.4 represents the percent of annual rainfall recorded during the respective time periods used for each station indicated in the previous columns. All daily values for the above time periods are divided by the total rainfall within the respective seasonal time period. The normalized values (as the daily percentage of the seasonal total) are numerically integrated (summed), leading to the respective ragged curves in Fig. 4.9: these are the observed data. This figure also shows the fit to these observations of the Cumulative Weibull function in (4.2), accomplished using iterative least squares until the rms misfit between the smooth model line and the observed data (the ragged line) stabilizes to four decimal places between iterations. Typical rms misfit errors are on the order of 1 % of the annual total, with an R2 coefficient of determination of 0.99.

Once the cumulative function $W(x, x_o)$ is determined, the corresponding probability density function $w(x, x_o)$ is determined through differentiation of $W(x, x_o)$, either numerically or analytically according to $w(x, x_o) = \partial W(x, x_o)/\partial x$. The resulting differentiated forms in Fig. 4.10 represent the expected daily percentage (%/day) of seasonal rainfall for each station. The data are read, for example, such that the maximum daily rainfall at KHA is typically expected on, or

Table 4.4 Range of daily values used to construct seasonal rainfall models

GED:	May 1–Oct 31	Calendar Day 120–300	(>98 % of annual prcp)
KAS:	May 1–Oct 31	Calendar Day 120–300	(>98.5 % of annual prcp)
KHA:	June 1–Oct 15	Calendar Day 152–273	(>92 % of annual prcp)

Fig. 4.10 Analytical model for the daily percent of annual rainfall for the three stations: GED, KAS, and KHA. The Weibull density functions derived from differentiating the forms in Fig. 4.9

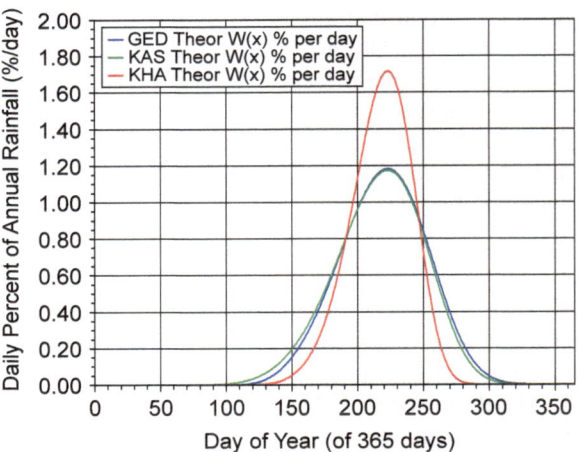

about, Day 223 (or Aug. 11), and will have a daily total of 1.7 % of the annual rainfall. The area under a partial segment of the curve in Fig. 4.10 will be the total rainfall expected during the respective time period. The latter information is also available from Fig. 4.9, where it might be noted that approximately 50 % of the annual rainfall at KHA is expected before Day 220 (Aug. 8). Thus the annual peak in daily rainfall at Khartoum generally coincides with the time when 50 % of the annual total has accumulated at or about the end of the first dekad in August (1 dekad = 10 days; there are approximately three dekads per month). The results in this Sect. 4.8, using a pre-conditioned seasonally trimmed database (see Table 4.4), generally accord with the unconditioned results in Fig. 4.8, which included data from earlier and later in the season.

During the early leading edge and late trailing edge of the monsoon season in Fig. 4.10, Kassala and Gedaref are relatively in phase, and have virtually identical seasonal morphologies within the resolution of the present data for the concurrent period of record. At the beginning of the rainy season, the start-of-season (SOS) at Khartoum lags Gedaref and Kassala by 8–10 days. This is in general accordance with the analysis using the full, untrimmed 12 months of daily values summarized in Fig. 4.8, although the analysis of the seasonal distribution of century-long sets of monthly values in Figs. 1.6 and 3.3 might imply that the monsoon season at Gedaref is somewhat longer than at Kassala. This difference may be due to the different periods of record used in the analyses.

In Fig. 4.1, the daily expected rainfall for Khartoum is significantly more seasonally peaked than for the other two stations, which is consistent with the prior analysis of long-term monthly values in this report (see Fig. 3.3). However, the seasonal rainfall at KHA drops off more quickly at the trailing edge relative to Gedaref and Kassala, with the end-of-season (EOS) leading GED and KAS by perhaps 15 days, and up to 20 days, depending on the criteria used to define the time of EOS.

References

Arkin PA, Meisner BN (1987) The relationship between large-scale convective rainfall and cold cloud over the Western Hemisphere during 1982–84. Mon Weather Rev 115:51–74

Balme M, Vischel T, Lebel T, Peugeot C, Galle S (2006) Sahelian water balance: impact of the mesoscale rainfall variability on runoff. Part 1: rainfall variability analysis. J Hydrol 331:336–348

CPC (2012) Climate Prediction Center, African Desk. http://www.cpc.ncep.noaa.gov/products/african_desk/cpc_intl/. Accessed 28 Sept 2012

CPC RFE 2.0 (2012) Africa RFE Technical Description. http://www.cpc.ncep.noaa.gov/products/fews/RFE2.0_tech.pdf. Accessed 26 Dec 2012

EarthInfo, 2010. EarthInfo, Inc. is a value-added vendor of WMO GHCN data. The product used here is Global Daily 2010. http://www.earthinfo.com/databases/gd.htm. Accessed 27 March 2013

El Tom MA (1975) The rains of the Sudan: mechanisms and distribution. Khartoum University Press, Khartoum

El Gamri T, Saeed AB, Abdalla AK (2009) Rainfall of the Sudan: characteristics and prediction. Arts J 27:18–35. Journal of the Faculty of Arts, Univ of Khartoum, Sudan. http://adabjournal.uofk.edu/current%20issue/ISSUES%20ENGLISH/El%20Gamri_%20Amir%2. Accessed 24 Feb 2013

GCOS (2004) Implementation Plan for the Global Observing System for Climate in Support of the UNFCCC, Executive Summary, October 2004, GCOS-92 (ES), (WMO/TD No. 1244). http://www.wmo.int/pages/prog/gcos/Publications/gcos-92_GIP_ES.pdf. Accessed 10 Oct 2012

GHCN-Daily (2012) The Global Historical Climatology Network, operated by the NOAA National Climate Data Center, provides historical daily data exchanged under the World Meteorological Organization (WMO) World Weather Watch Program. http://www.ncdc.noaa.gov/oa/climate/ghcn-daily/. Accessed 20 Oct 2012

Hermance JF, Sulieman HM (2013) Comparing satellite RFE data with surface gauges for 2012 extreme storms in African East Sahel. Remote Sens Lett 4(7): 696–705. doi: 10.1080/2150704X.2013.787498

Lebel T, Ali A (2009) Recent trends in the Central and Western Sahel rainfall regime (1990–2007). J Hydro 375:52–64

Gebremichael M, Hossain F (eds) (2010) Satellite rainfall applications for surface hydrology. Springer, Heidelberg

Weibull W (1951) A statistical distribution function of wide applicability. J Appl Mech-Trans ASME 18(3):293–297. http://www.barringer1.com/wa_files/Weibull-ASME-Paper-1951.pdf. Accessed 27 April 2012

WMO Normals (2012) ftp://dossier.ogp.noaa.gov/GCOS/WMO-Normals/RA-I/SU/. Accessed 24 Nov 2012

Xie P, Arkin PA (1996) Analysis of global monthly precipitation using gauge observations, satellite estimates, and numerical model prediction. J Clim 9:840–858

Chapter 5
Analysis of Storm Events (and Interstorm Dry Periods)

Abstract In the East Sahel, it is important to know whether monsoon rains occur over extended periods of contiguous days or are episodic in nature. Not only are the statistics of storm duration important for proper water management, but also statistics on the intervals between storms—intraseasonal and interannual dry periods. Storm statistics reported here for three study sites in the East Sahel accord well with results reported by others for elsewhere in the Sahel. There are typically 33 storm events annually at Gedaref (GED), 20 events at Kassala (KAS), and 11 events at Khartoum (KHA); their respective mean storm totals are 18 mm (GED), 13 mm (KAS) and 12 mm (KHA). Statistical metrics on storm total and duration are derived from the percentage of the annual number of storms and from the percentage of the annual total rainfall that each storm contributes. Much like what is reported for the West Sahel, there are significantly more smaller events than larger ones at these three East Sahel stations; however, it is the fewer, larger storms that deliver the greater percentage of the annual total rainfall. Storm totals greater than 50 mm are an important class of extreme events, with approximately 5 % of the storm events at these stations falling into this category. Such extreme events tend to occur between Day 150 and Day 300 of the calendar year. Although there appears to be little tendency for storms to cluster at particular times of the season, their temporal center of mass tends to fall in the July to August sector. Interstorm dry periods during the monsoon or growing season commonly exceed 20 days. Interannual dry periods between the monsoons at GED exceed 200 days approximately 25 % of the time (one out of 4 years), 230 days at KAS, and 250 days at KHA.

5.1 Statistics of Storm Events

A primary question of interest is: Does rain during the monsoon season occur over a continuum of many days—such as a rainy *season*—or is it more episodic? The answer is fundamental to understanding the response of natural systems to rainfall

J. F. Hermance, *Historical Variability of Rainfall in the African East Sahel of Sudan*, SpringerBriefs in Earth Sciences, DOI: 10.1007/978-3-319-00575-1_5, © The Author(s) 2014

as well as for proper water management. The storm events discussed in this chapter are associated with the northward seasonal march of the African Monsoon, the tropical rainbelt and the Intertropical Convergence Zone (ITCZ)—a topic intensively studied in West Africa (Nicholson 2009; D'Amato and Lebel 1998; Bell and Lamb 2006). While this topic is relatively under-studied in the East Sahel—in the region of the three long-term stations available for the work reported here—this report will show that much of the general morphology (the spatial and temporal patterns) of rainfall identified in the West Sahel seems to apply to the East Sahel. This is particularly true of the dominant role that deep convective uplift plays in rainfall variability across the Sahel (Guy and Rutledge 2012). Convective storms are the primary source of rainfall in the East Sahel (El Tom 1975), often having significant impacts on local communities and economies as extreme events (Sutcliffe et al. 1989; Hulme and Trilsbach 1989).

5.1.1 Defining Storm Events

For the purpose of this discussion, a storm "event" will be defined as a specific episode of recorded (i.e. *measurable*) precipitation having a continuous duration of one or more days. In addition, the total rainfall for any event (regardless of duration) must equal or exceed 1 mm. All candidate events having totals less than 1 mm are simply dropped from the analysis. For example, if a measurable amount of rain is recorded on each of three contiguous days and the total of daily values for this interval is 1 mm or greater, this is classified as a 3-day storm event. Such an event will be characterized by *duration* (in days) and by storm *total* (or *magnitude*), namely the total amount or depth of rainfall (in mm) that falls over the duration of the storm. The term "intensity" is often used by hydrologists and meteorologists (and is used in other chapters in this text), and refers to rainfall *rate*—depth of accumulation per unit time—but is used with particular caution in this chapter, since it is quite unlikely that the true instantaneous intensity of rainfall during a three-day storm can be represented by the storm total divided by the number of days. Studies elsewhere in the Sahel imply that what is referred to as a one-day storm "event" might in reality be a short—often lasting an hour or less—downpour (e.g. D'Amato and Lebel 1998; Balme et al. 2006). Events of more than a day in this study are likely to be a series of short, intense downpours on contiguous days. Unfortunately, only daily data are available from the Global Historical Climate Network (GHCN-Daily 2012) for the present study area. Intensities therefore will be reported as rainfall totals per day (mm/d), and over the duration of a multiday storm event, storm totals or magnitudes will be reported in units of depth (mm).

5.1.2 Partitioning Storm Event Totals as a Percentage of Occurrence

The first question to be addressed is: Given that a storm occurs—regardless of duration—what is the probability that the total rainfall for the event will exceed a specified value? A procedure for assessing this statistic is illustrated in Fig. 5.1 for Gedaref (GED). Storm days have been collected into storm events. Assume that n is the total number of discrete events, each associated with a specific total rainfall (and duration, although that does not enter the picture yet). The collection of events is then ranked by *decreasing* (or, alternatively, increasing) magnitude or storm total, and assigned a rank index k, where $k = 1$ for the *largest* (or smallest) event, and $k = n$ for the smallest (or largest). Each member of the collection is now assigned a Weibull statistic $k/(n + 1)$. Figure 5.1 uses this Weibull statistic on a series of ranked *decreasing* values to determine the probability that a storm event will *exceed* the total rainfall indicted along the abscissa. This statistic in Fig 5.1 is referred to as the fractional percent frequency of exceedence, or, elsewhere in this report as *exceedence* probability $E(x)$.

If, instead of the procedure of ranking data in *decreasing* order, as in Fig 5.1, the storm event totals were ranked by *increasing* values, the Weibull statistic would provide a measure of the *cumulative* probability usually denoted by the expression $P(x)$. This is the probability that a storm event will be *less than* the

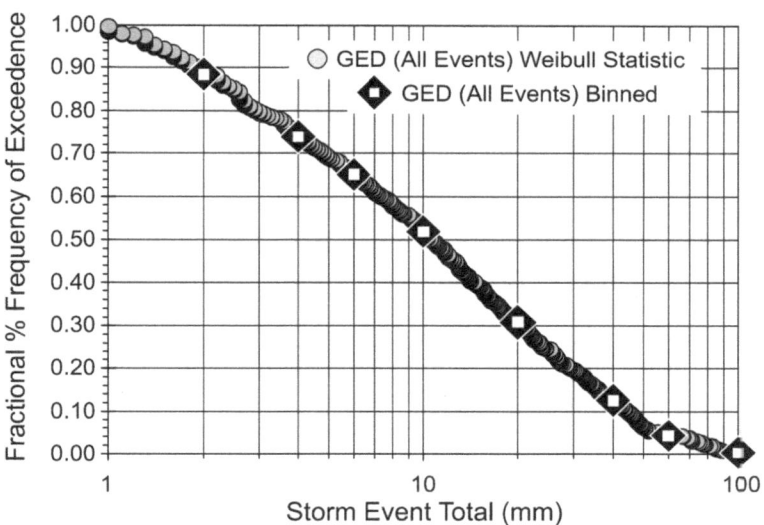

Fig. 5.1 Exceedence probability of storm totals for all storm events at Gedaref. (The ranked Weibull statistic is compared with the cumulative binned statistics.) Approximately 50 % of the storms exceed 10 mm rainfall; approximately 10 % of the storms exceed 44 mm

assigned value along the abscissa. The cumulative probability $P(x)$ is, of course, related to the exceedence probability by $E(x) = [1.0 - P(x)]$.

Shown in Fig. 5.1 (by the diamond symbols) is an alternative procedure that assesses the magnitude of each storm event, then apportions it into one of a number of contiguous bins of increasing values. The number of events in each bin is normalized by the total number of storm events. The discrete bin values— percentages of the total number per bin—are then integrated (summed) from bin to bin into a cumulative statistic $P(x_i)$ defined at the respective edge x_i of each bin. Using the relation $E(x_i) = 1.0 - P(x_i)$, the binned values are aggregated into another form of *exceedence probability*: the probability that a storm event will exceed the upper edge of the respective bin as shown by the diamond symbols in Fig. 5.1. For the applications here, the two procedures generally agree to better than the order of 10 %.

As an example of interpreting these curves, both with respect to the Weibull ranking and with respect to the binning procedure, the median rainfall for a storm at Gedaref, according to Fig. 5.1, is approximately 10 mm. Figure 5.1 indicates the probability that a storm total will be greater than 20 mm at Gedaref as approximately 30 %. There is a 5 % possibility that, if a storm occurs, it will exceed a total rainfall of 50 mm. At the other extreme, 90 % of the storms have totals greater than 2 mm. (Regarding the latter statement, however, the interpreter of these statistics needs to keep in mind that events having storm totals of less than 1 mm have been eliminated from the analysis.)

Cumulative statistics for all three stations are compared in Fig. 5.2 (the GED curve in this figure is a reformatted version of the data in Fig. 5.1).

Data have been binned into quasi-geometrically increasing intervals (with bin edges at 0, 1, 2, 5, 10, 20, 35, 50, 75, etc.) from zero to the maximum storm total for each site. This is basically a smoothing procedure, with the statistics being most representative at the bin edges (at the symbols in the figure), while the linear interpolation between the edges is more approximate. The interpolated values nominally agree with the original observed data to within 5 %, but one should not assume a better agreement than 10 %. It should also be noted that a few pilot analyses that included events with storm totals lower than 1 mm significantly altered the probability distribution at the low end of the values, so one should keep in mind that event totals of less than 1 mm have been clipped—dropped—from the analysis.

Comparing the results from the three stations in Fig. 5.2, consider for example that approximately 50 % of the storm event totals at Gedaref (GED) are less than 10 mm, whereas at Kassala (KAS) and Khartoum (KHA) 60 % of the storm events have totals less than 10 mm. The median storm-event rainfall at KAS and KHA is approximately 6 mm. There is approximately a 70 % probability that the total rainfall for an event at GED will be less than 20 mm, and an 80 % probability that events at KAS and KHA will be less than 20 mm. The cumulative distributions for generic storm events at Kassala and Khartoum are remarkably similar, whereas the distribution at Gedaref seems to reflect the more humid setting of that station.

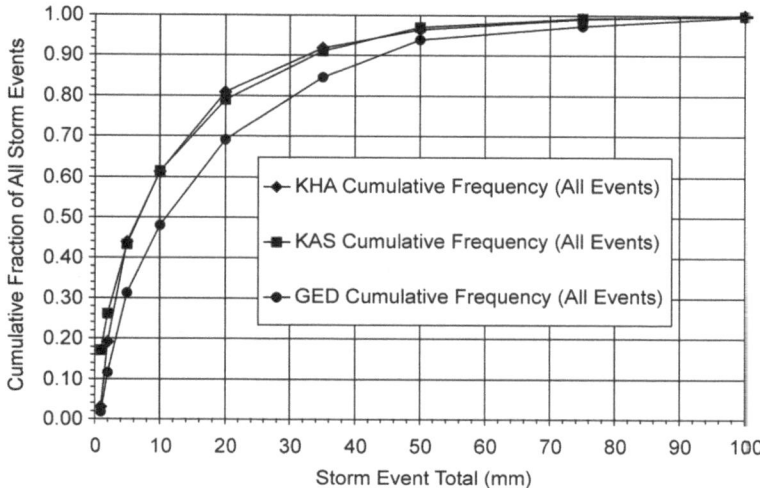

Fig. 5.2 Percentages (%) of annual storm events contributed by events having totals less than the value of the respective abscissa. For example, for *GED*, the number of annual events from storms having totals less than 20 mm is approximately 70 %

5.1.3 Partitioning Storm Event Totals by Duration

Figures 5.3, 5.4, and 5.5 apportion individual storms recorded at each station by their respective duration. Numerical details on all events for all three stations are provided in Table 5.1. As seen in the table, most of the storms have durations of 3 days or less, so that only the analyses for storms lasting 3 days or less are summarized in the figures. Two types of statistics are summarized in each figure. First, there is the cumulative fraction of storm events at the respective station that have total rainfall less than the storm event total along the abscissa (*x*-axis). These (the "All Events" category referred to the left axis) are simply reproduced from the corresponding curves in Fig. 5.2. The second type of statistic involves sorting the storm events into their respective durations (1-day, 2-day, etc.), then determining the percentage (%) of the number of storms below a certain threshold (in mm) that is comprised of storms of the specified duration (the "class"). The associated percentage (%) of all storms in the respective class (duration) having rainfall totals less than the value along the abscissa is referred to the vertical axis on the right.

As an example, for the case of Gedaref (GED) in Fig. 5.3, the cumulative fraction of all storms having rainfalls of 20 mm or less is 0.70 (70 %). A fraction—0.80, or 80 %—of these will be 1-day storms; a fraction of 0.15, or 15 %, will be 2-day storms; and a fraction of about 0.04, or 4 %, will be 3-day storms. These three classes do not account for the full 100 % of the storm events because the contributions of 4-day, 5-day and 6-day storms are not considered.

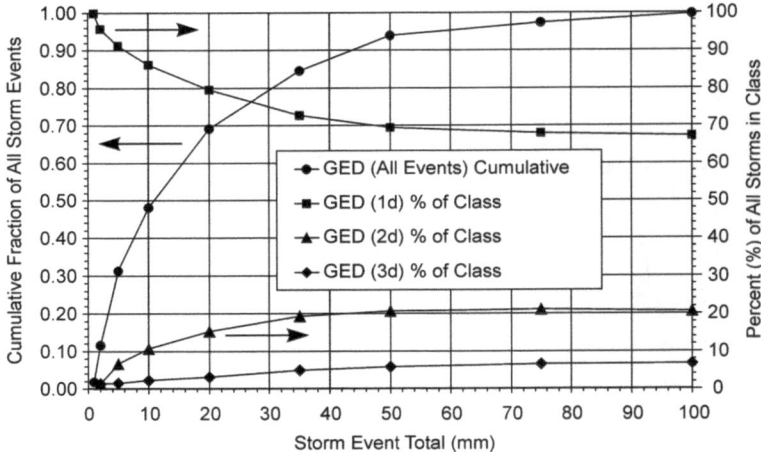

Fig. 5.3 Storm event magnitude-duration statistics for Gedaref. Classification of the expectation of storm events by event total (or magnitude) and duration. *Left axis* Cumulative fraction of storm events less than the specified value on the x-axis. *Right axis* Fraction of storm events, less than the specified total, having the indicated duration. For example, 70 % of all storms will have total rainfalls of 20 mm or less; of these 80 % will have durations of 1 day

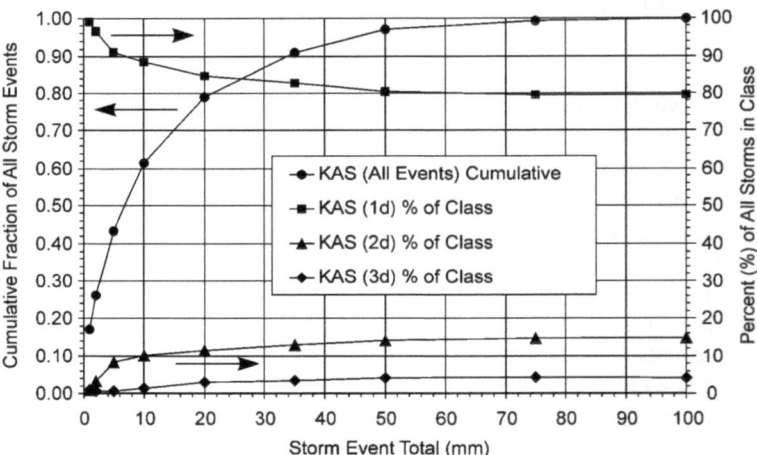

Fig. 5.4 Storm event magnitude-duration statistics for Kassala. See Fig. 5.3 caption for explanation

As a side note, one needs to be aware of the possibility of aliasing the contributions from n−day events into $(n + 1)$−day events. For example, if a portion of daily values recorded during what is a 2×24 h $= 48$ h interval overlaps onto a third recording day. If the storm starts at noon on Day 1 and runs to noon on Day 3,

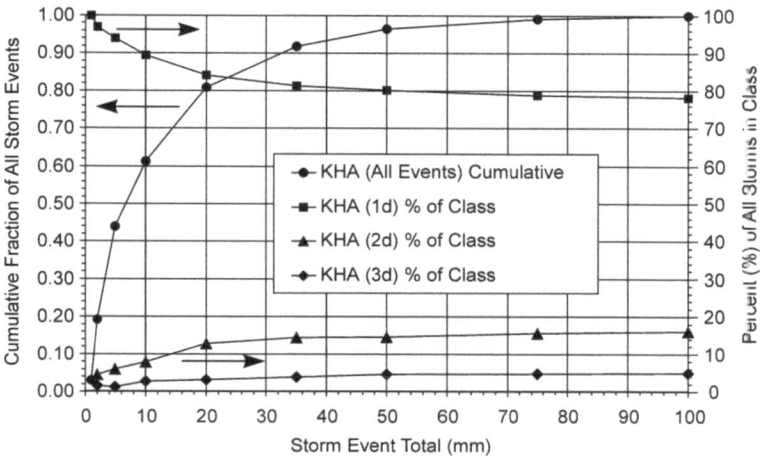

Fig. 5.5 Storm event magnitude-duration statistics for Khartoum. See Fig. 5.3 caption for explanation

this 48 h storm would be tagged as a 3-day storm event. Clearly this would lead to a mixing of the duration statistics, resulting in an underestimate of the number of n-day events, and an overestimate of the number of $(n + 1)$−day events.

The storm-duration and storm-total characteristics for Kassala in Fig. 5.4, indicate a median rainfall of approximately 6 mm for storm events. In addition, 80 % of the storms will have rainfall totals of 20 mm or less, and of those, 85 % will have durations of 1 day.

Results for the driest station, Khartoum, are shown in Fig. 5.5. Similar to the figures for Kassala, approximately 80 % of all storms have totals of less than 20 mm, Moreover, the partitioning of the duration of storms accounting for the total number at Khartoum is remarkably similar to that for Kassala.

5.1.4 Some Numerical Details on Storm Duration and Totals

Additional details on storm events for the three stations are summarized in Table 5.1. The respective periods of record are given in the table captions, with the coverage being 99 % for Gedaref, 98 % for Kassala, and 99 % for Khartoum. The days with recorded data provide the number of daily data samples for the analysis; days with no data have been excluded from the latter totals; precipitation estimates for days with recorded "trace" amounts are set to zero. "Storm days" in Table 5.1 are those days associated with storm events of one day or longer, as defined earlier in this chapter. In cases where the number of storm events per year is less than one, such as $n = 0.2$ for the 6-day storm events for Gedaref in Table 5.1a, it may be best to think of these fractions as the inverse of the return period T where

Table 5.1 Gedaref: summary statistics for daily precipitation

a. Gedaref: (1957–1975; n = 6,869 daily samples)

Storm duration (days)	1	2	3	4	5	6	All classes
Total "Storm Days" of record	417	256	123	88	35	18	944
Total "Storm Days" per year	22	13	6	5	2	1	50
Number of storm "Events" of record	417	128	41	22	7	3	621
Number of storm "Events" per year	22	7	2	1	0.4	0.2	33
Percentage of storm "Events" of record (%)	67	21	7	4	1	0.5	100
Mean storm total (mm)	13	23	33	43	68	38	18
Median storm total (mm)	7	20	27	34	83	32	11
Median storm intensity (mm/d)	7	10	9	8	16	5	8
Total precipitation all events in class (mm)	5,239	2,943	1,350	954	475	113	1,1078
Total annual prcp all events in class (mm)	276	155	71	50	25	6	583
Percent (%) of all precipitation	47	27	12	9	4	1	100
Cumulative distribution (% equal to or less than)	47	74	86	95	99	100	

b. Kassala: (1957–1990; n = 12,179 daily samples)

Storm duration (days)	1	2	3	4	5–6	All classes
Total "Storm Days" of record	548	206	87	28	11	882
Total "Storm Days" per year	16	6	3	1	0.3	26
Number of storm "Events" of record	548	103	29	7	2	690
Number of Storm "Events" per year	16	3	1	0.2	0.1	20
Percentage of Storm "Events" of record (%)	79	15	4	1	0.3	100
Mean storm total (mm)	10	22	24	33	27	13
Median storm total (mm)	5	16	19	31	27	7
Median storm intensity (mm/d)	5	8	6	8	5	6
Total precipitation all events in class (mm)	5,650	2,209	687	229	53	8,829
Total annual prcp all events in class (mm)	166	65	20	7	2	260
Percent (%) of all precipitation	64	25	8	3	1	100
Cumulative distribution (% equal to or less than)	64	89	97	99	100	

c. Khartoum: (1957–1987; n = 11,173 daily samples)

Storm duration (days)	1	2	3	4	All classes
Total "Storm Days" of record	268	110	51	12	441
Total "Storm Days" per year	9	4	2	0.4	14
Number of storm "Events" of record	268	55	17	3	343
Number of storm "Events" per year	9	2	1	0.1	11
Percentage of storm "Events" of record (%)	78	16	5	1	100
Mean storm total (mm)	9	22	27	23	12
Median storm total (mm)	4	15	21	24	7
Median storm intensity (mm/d)	4	8	7	6	6
Total precipitation all events in class (mm)	2,445	1,211	465	68	4,190
Total annual prcp all events in class (mm)	79	39	15	2	135
Percent (%) of all precipitation	58	29	11	2	100
Cumulative distribution (% equal to or less than)	58	87	98	100	

$T = 5$ years. For some of these infrequent events, of course, the number of samples may not warrant a great confidence in the respective statistic, and should be treated as a provisional, or qualitative, estimate for planning. For example, there were only three storm events at Gedaref of 6 days' duration for the 19-year period of record. In the case of Kassala, there was only a single 5-day event and a single 6-day event, which are here combined for a single composite 5–6 day event entry. While highly unreliable, the statistics for the long-duration events indicate how exceptional it is for the East Sahel to experience storm systems lasting more than several days.

5.1.5 Partitioning Storm Event Totals as a Percentage of Annual Rainfall

As an alternative to using the percentage of the total number of events as the base for assigning the statistic, as in the last section, the analysis in this section will employ the percentage of annual rainfall that is contributed by a given class of storm events. First, it should be noted in Table 5.1 that the metric "Total annual prcp all events in class (mm)" for "all classes" is a measure of the total annual rainfall for each respective station. Each subsidiary value, however, does not agree precisely with its corresponding value based on the long-term annual composites of monthly values in Chap. 1, Table 1.1, nor with values for the annual normals cited in Chap. 2, Table 2.1. In the case of Gedaref, the annual storm totals for the 19-year period of record (1957–1975) have a mean total of 583 mm, whereas the long term expected value (LTEV) in Table 1.1 equals 600 mm. For Kassala, the annual storm totals have a mean of 260 mm, where the LTEV = 280 mm; and for Khartoum, the storm totals have a mean of 135 mm, where the LTEV = 130 mm. Actually, the values agree relatively well, considering that different periods of record are involved, the respective means as opposed to the long-term medians are compared, and for the storm event analyses in this chapter, numerous events having rainfall less than 1 mm have been ignored.

But in order to compare patterns of precipitation—rather than absolute totals—among stations, in this section the magnitude of each discrete storm event is normalized by the sum total of precipitation from all storms for the respective station. Selected numerical parameters determined from the present analysis are summarized in Table 5.1, along with those from Sect. 5.1.3.

The cumulative percentage of annual rainfall at each station, contributed by all storm events up to and including a specified storm total—regardless of duration—is shown in Fig. 5.6. For example, according to Fig. 5.6, in the case of Gedaref (GED), storms of magnitude 20 mm or less contribute about 30 % of the annual rainfall. However, in the previous Sect. 5.1.3, according to Fig. 5.3, the cumulative fraction of all storms at Gedaref having rainfalls of 20 mm or less is approximately 70 %. Merging these two observations, we find that 70 % of the storm events are less than 20 mm, yet events less than 20 mm contribute only 30 % of the annual rainfall. If 30 % of the annual rainfall is from events measuring less than 20 mm, couched in terms of exceedence probability, this implies that 70 % of the annual rainfall is contributed by events greater than 20 mm.

According to Fig. 5.6, the cumulative contributions of storm totals are virtually identical for Khartoum and Kassala as compared to Gedaref. For these two stations, 20 % of the rainfall is from storms of less than 10 mm, which is to say 80 % of the rain is contributed by events larger than 10 mm. According to Fig. 5.2, only 5 % of the storm events are larger than 50 mm at KHA and KAS, yet according to Fig 5.5, these larger storms (storm prcp > 50 mm) account for 15–20 % of the annual rainfall for these two stations. In general terms, 80–90 % of the annual rainfall at all three stations is contributed by storm events having totals greater than 10 mm, whereas according to Fig. 5.2, only 40 % (KAS & KHA) to 50 % (GED) of the actual storms have totals greater than 10 mm.

Fig. 5.6 Cumulative percentage of annual rainfall contributed by all storm events less than or equal to the indicated storm event total. Results shown for all storm durations at Gedaref (*GED*), Kassala (*KAS*) and Khartoum (*KHA*)

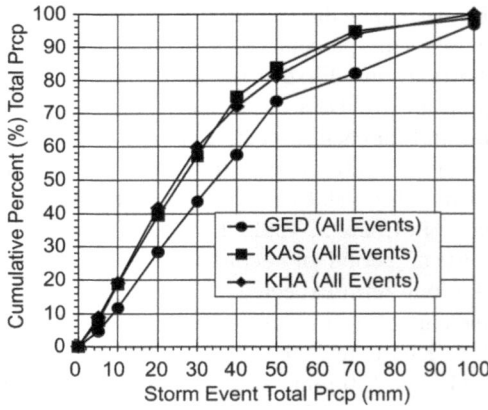

Fig. 5.7 Percentage of annual rainfall contributed by storm events exceeding the indicated storm event total: Gedaref, Sudan

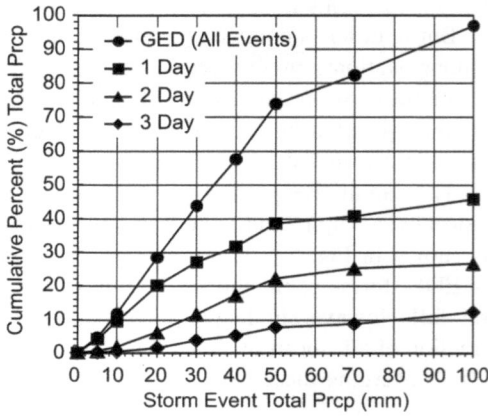

Fig. 5.8 Percentage of annual rainfall contributed by storm events exceeding the indicated storm event total: Kassala, Sudan

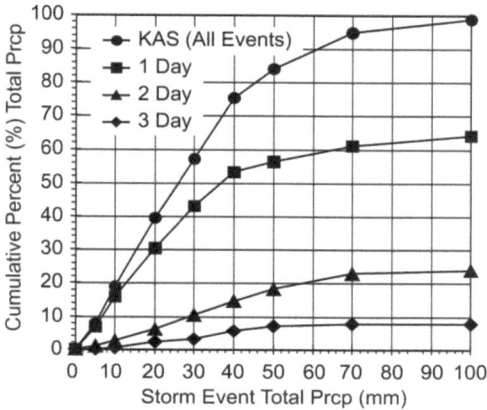

Fig. 5.9 Percentage of annual rainfall contributed by storm events exceeding the indicated storm event total: Khartoum, Sudan

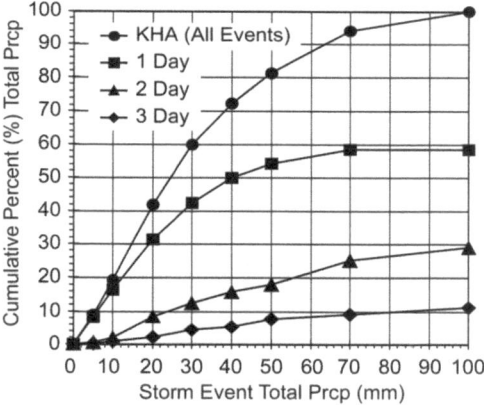

These statistics are broken down by duration in Fig. 5.7 for Gedaref, in Fig. 5.8 for Kassala, and in Fig. 5.9 for Khartoum, with additional summary information in Table 5.1. Storm events of durations up to 3 days are included in these plots, since the statistics on the storms of longer duration are less robust because their number is far fewer. Indeed, even for the plots shown, the results for events having rainfall totals greater than 50 mm ought to be used with some caution.

According to Fig. 5.7 for Gedaref, Fig. 5.8 for Kassala, and Fig. 5.9 for Khartoum, and as summarized in Table 5.1, storm events of 1-day duration generally contribute the highest percentage of annual rainfall at all three stations.

The asymptotic cumulative values to the right of each graph for each station and duration accord with the values listed in the second line from the bottom of the table for the respective station in Table 5.1. Note the relative similarity among corresponding durations for the respective stations, even though their expected annual totals differ widely.

5.1.6 Essential Characteristics of Storm Events in the East Sahel

5.1 6.1 Extreme Events: Statistics on Duration and Magnitude

For purposes of this discussion, an "extreme event" will be provisionally defined as any storm having less than a 10 % probability of occurrence. This corresponds with the far-right-hand section of Figs. 5.3, 5.4 and 5.5. The statistics of the numbers of such events are summarized in Table 5.2. The total number of all extremes is not particularly meaningful in itself, unless one divides that number by, for example, the total number of years of record for the respective gauge. However, the total of all extremes is a variate used to compute the statistic for the percentage of extreme values having the indicated duration. Clearly, there are proportionately more extreme 1-day events at Kassala and Khartoum than there are at Gedaref. As for other metrics in this chapter, the partitioning of these extreme events reveals greater similarity between Kassala and Khartoum, than it does between either of these stations and Gedaref.

Table 5.3 summarizes the range of storm totals (or magnitudes) for extreme events at each station. Two metrics are provided. One is the respective storm total at the 1 % exceedence probability—for example, in Table 5.3, 1 % of the values at

Table 5.2 Statistics on the numbers and duration of extreme storm events

	Gedaref		Kassala		Khartoum	
Storm duration (days)	Number of events	Percentage of extreme values	Number of events	Percentage of extreme values	Number of events	Percentage of extreme values
6	1	1.5	1	1.4	–	–
5	5	7.7	–	–	–	–
4	9	13.8	3	4.3	1	2.4
3	12	18.5	9	12.9	6	14.6
2	17	26.2	24	34.3	15	36.6
1	21	32.3	33	47.1	19	46.3
Total number of all extremes	65	100	70	100	41	100

Table 5.3 Range of extreme storm total precipitation

Station	Exceedence probability (All storms)	
	1 % (mm)	10 % (mm)
GED	85	44
KAS	65	34
KHA	66	27

Gedaref are expected to have storm totals greater than 85 mm. The other metric is the respective storm total at the 10 % exceedence probability, the definition of extreme events. For Gedaref, for example, any storm greater than 44 mm would be classified as an extreme event; at Kassala, an extreme event would be greater than 34 mm.

Note that, in keeping with the discussion elsewhere in this report, although total annual rainfall at Khartoum is significantly less than that at the other two stations, Table 5.3 indicates that rainfall totals from extreme storms follow a different proportion. KHA, for example, has less than 25 % of the annual rainfall at GED (see Chap. 1); however, the most extreme storm totals at Khartoum (say, at the 1 % exceedence level) are only 78 % of those at Gedaref (66 mm/85 mm).

For practical purposes, in terms of their potential for causing local flash flooding, one might consider storms having totals greater than 50 mm as an important class of extreme events. Approximately 5 % of the storm events at these stations fall into this category (6 % at GED, and 3–4 % at KAS and KHA). With respect to annual totals, 25 % of the annual rainfall at GED is contributed by events greater than 50 mm, and approximately 15–20 % of the annual rainfall at KAS and KHA.

5.1.6.2 Temporal Distribution of Storm Totals

The timing and magnitude of these storm events over a season, for a number of years, is illustrated in Fig. 5.10, where storm events for a period of record common to the three stations have been sorted by day of the year and by year of record. The size of the symbols is scaled to be proportional to each storm total, following the key in the lower right corner of the bottom panel.

One of the first things to notice is the significant increase in the density (number per year) of events as one inspects the station data from north (top panel) to south (bottom panel), or from typically dry conditions to wetter conditions. Not surprising (as indicated in the discussion of the seasonal distribution of monthly rainfall elsewhere in this report) is the strong tendency for significant storms to occur between Day 150 and Day 300, with the storms at Khartoum possibly occurring over a shorter season. Within the rainy season at each station, there seems to be little tendency, in the time frame of these data, for extreme storms to cluster later or earlier in the season. For example, Fig. 5.10 shows several anomalous events for Gedaref (GED)—one in 1958, and the other in 1966. The maximum daily total for the 1958 event was 172 mm, and it occurred in September, whereas the 1966 event occurred in June and had a maximum daily total of 89 mm. Hence, both events lie outside the typical peak rainfall months for the monsoon season—July and August—although the seasonal center of mass for all the storm events tends to fall in the July-to-August sector.

Storms like the two extreme events for Gedaref in Fig. 5.10 are essentially outliers, and too infrequent to categorize statistically. Ignoring such extremes in Fig. 5.10, the pattern of the remaining members would qualitatively suggest that the

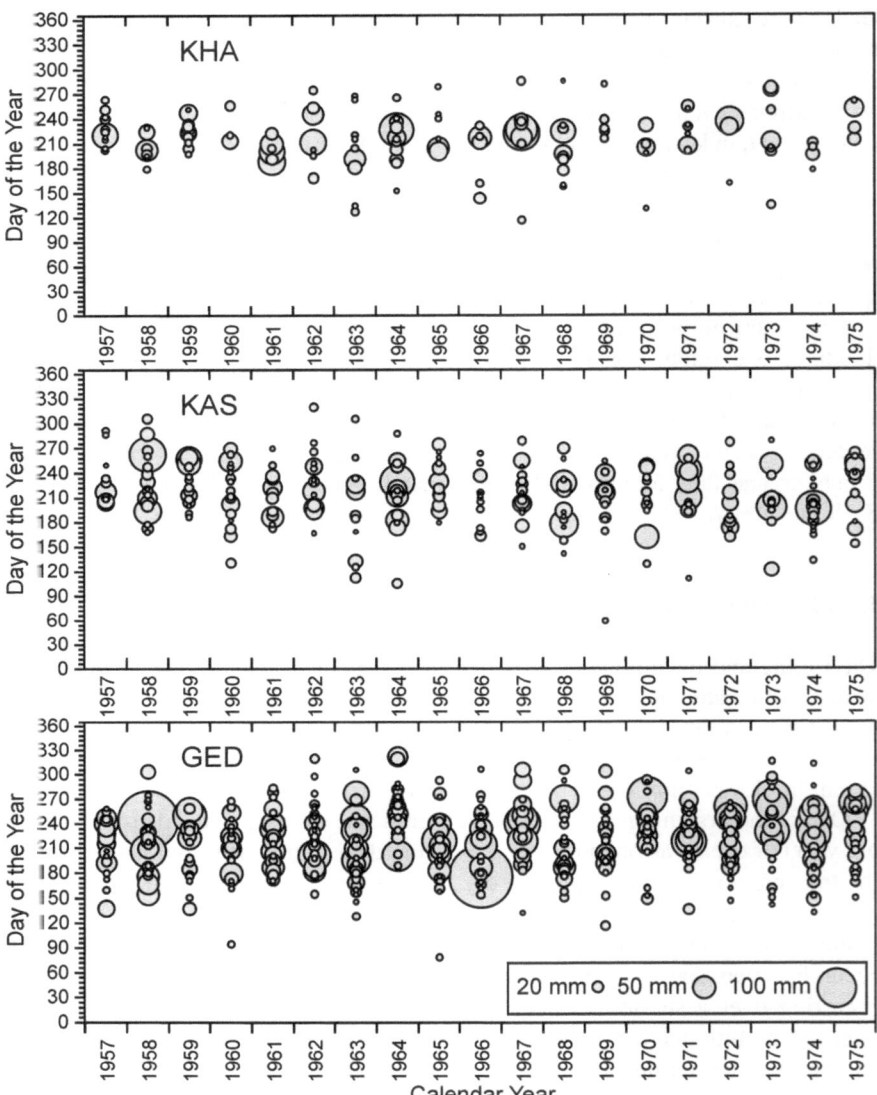

Fig. 5.10 Temporal distribution of storm event totals by year and season

principal difference among the three datasets appears to be a difference in the *number* of events at each site, rather than any systematic difference in the overall *magnitude* of storm events. This is explored more quantitatively in the next section.

5.1.6.3 Summary of Results on Storm Event Magnitudes and Durations

One of the important questions about the behavior of monsoons in this area of the East Sahel is the duration of typical storms. It appears that extended periods of storm activity over many contiguous days are unusual. For example, for Gedaref in Table 5.1a, note that 86 % of all storms have durations of 3 days or less, and 95 % have durations of 4 days or less. No storm lasting more than 6 days has been recorded over the 19-year period of record from Gedaref. At Kassala, no storm lasting longer than 6 days has ever been recorded over its 34-year period of record. In fact, at Kassala, only one 5-day event and one 6-day event have been recorded over these 34 years. At Khartoum (a 31-year period of record), there were no storms of duration greater than 4 days, and only 3 events of duration longer than 3 days. As a class, 1-day storm events predominate in the weather record over those of longer duration. As summarized in Table 5.1, 1-day events at Gedaref account for 67 % of the storms. At Kassala, they account for 79 % of the storms; and at Khartoum, they account for 78 % of the storms. This is consistent with the type of localized, short-term convective storm systems responsible for the majority of rainfall events in the East Sahel (El Tom 1975).

The analysis in previous sections of this chapter explored patterns of storm event totals (or magnitudes) versus storm event numbers (or frequency), and the percentage of annual rainfall delivered by the magnitude of storms. As an example, consider the aggregate of all storm events at Gedaref. Figure 5.3 shows that 70 % of Gedaref storm events have totals of *less* than 20 mm, whereas Fig. 5.7, based on total rainfall, shows that 70 % of the annual precipitation at this station is contributed by events with magnitudes *greater* than 20 mm. In other words, while there are significantly more smaller events than larger ones, it is the fewer, larger storms that deliver a proportionately greater percentage of the annual rainfall. This pattern applies to the storm event data from all three reference stations and accords with patterns reported by a number of workers in the West Sahel (Lebel and Ali 2009).

A related issue is the actual magnitude of storm events contributing to the significantly different rainfall totals for each region. Recall from Chap. 1 that the long-term (century-long) median annual precipitation is 130 mm at Khartoum (1899–2010), 280 mm at Kassala (1901–2010), and 600 mm at Gedaref (1903–2010). Annual rainfall at Khartoum is approximately 22 % of that at Gedaref, smaller by almost a factor of 5; and annual rainfall at Kassala is approximately 47 % that of Gedaref, smaller by a factor of more than 2. Are these differences largely due to differences in the magnitude of discrete events, or to differences in the number of events at each site?

The first step is to verify that the daily data from the stations' more limited period of record are consistent with the monthly and annual values from their century-long period of record reported in Chap. 1, Table 1.1. Table 5.4 compares, for each station (column 1), its long-term median value computed from the century-long monthly data (column 2) with its mean annual total computed from the dai_y data (column 3), which have a nominal period of record of from 2 to 3 decades. Also shown in parentheses are the ratios of the reported annual precipitation at each station relative to that at Gedaref, the southernmost, wettest station. The mean annual total from the daily values (column 3) are computed using the data in Table 5.1. First determined is the total rainfall for all events from the period of record—for GED, for example, this total was 11078 mm (see Table 5.1a). This total is then divided by the number of years of record. GED, for example, has 19 years of record, so that the mean annual total is 583 mm. As mentioned in Sect. 5.1.4, the values in column 2 and column 3 agree relatively we l, considering the different procedures employed. And although the annual totals in column 3 computed from the daily values are somewhat smaller than the long-term median values in column 2, the ratios in column 3 with respect to Gedaref show the same general pattern as those in column 2, as discussed earlier in this section. The total annual rainfall at Kassala is 44–47 % (approximately half) that at Gedaref, and the total annual rainfall at Khartoum is 22–23 % (approximately a quarter) that at Gedaref, so that the rainfall totals from the daily data are consistent with those from the long-term expectation values.

The analysis in this chapter has already shown that the greatest percentage of annual rainfall is contributed by the larger storms at each site, so the key question becomes: Is there a systematic difference in the statistical distribution of storm event magnitudes among the respective stations? Table 5.5 compares numerical values for the three stations picked directly from the curves for the cumulative percentages in Fig. 5.6. The table shows the magnitudes of storms responsible for the specified percentage of annual rainfall and compares the corresponding magnitudes at KAS and KHA with those at GED.

The first entry in the table, for example, shows that 20 % of the annual precipitation at Gedaref is caused by storm events precipitating less than 15 mm. For Kassala, 20 % of the annual precipitation is caused by storm events of less than

Table 5.4 Compare typical storm event magnitudes between stations with annual totals

Station	Annual total (Monthly)[a]	Annual total (Daily)[a]
GED	600 mm (1.00[b])	583 mm (1.00[b])
KAS	280 mm (0.47)	260 mm (0.44)
KHA	139 mm (0.22)	135 mm (0.23)

[a] Annual total for respective period of record of data
[b] Ratio with respect to GED value

Table 5.5 Compare cumulative distributions of percent (%) of annual precipitation

Station	20 % [a] (mm)	Ratio[b]	50 % [a] (mm)	Ratio[b]	80 % [a] (mm)	Ratio[b]
GED	15	1.00	34	1.00	66	100
KAS	11	0.73	26	0.76	47	0.71
KHA	10	0.67	24	0.71	50	0.76

[a] Cumulative probability (see Fig. 5.6)
[b] Ratio with respect to GED value

11 mm. Note how similar the value at Kassala (11 mm) is to that at Khartoum (10 mm), particularly considering that there is almost a factor of 2 difference between their annual totals. The similarity between Kassala and Khartoum carries across all classes in Table 5.5. In addition, the table shows that the expected storm-event magnitudes for both Kassala and Khartoum contributing to their respective annual totals are only 67–76 % of those at Gedaref, whereas Gedaref has an annual total rainfall twice that of Kassala, and more than four times that of Khartoum. Thus, the differences in the annual totals at each station are not primarily due to relative differences in storm magnitudes, but rather to differences in storm frequency (Table 5.1). For example, according to Table 5.1, there are typically 33 annual events for GED, 20 such events for KAS, and 11 such events at KHA: a distribution significantly more representative of the distribution of annual totals in Table 5.4. However, the concordance is even stronger between the distribution of annual totals for the three stations and the number of storm or rain *days* per year (Chap. 4). As shown in Table 4.2, the typical apportioning of rain days per year for GED, KAS and KHA is, respectively, 56 day/year, 30 day/year, and 15 day/year. Hence, the total annual rainfall at Kassala is 44–47 % of that at Gedaref, and the number of rain days at Kassala is 53 % of those at Gedaref. Hence, the total annual rainfall at Khartoum is 22–23 % of that at Gedaref, while the number of rain days at Khartoum is 26 % of the number at Gedaref. The total annual rainfall tracks very well with the number of rain days.

5.2   Time Interval Between Storms: Dry Periods

5.2.1   Statistics of Interstorm Intervals

Another climate variable of some interest to applied science and to water managers is the time interval between storms (hereafter, the *interstorm interval*) defined here as the number of days between discrete storm events, where more than a trace amount of rainfall occurs on both the last day of the last event and the first day of the following event. Statistics are computed using the complete period of record for the respective station. Assessing the characteristics of such a parameter for a local area is important in mitigating the impact of interannual and intraseasonal dry periods and droughts, and in particular to the design of water storage systems. The scale and timing of interstorm intervals for the three reference stations are schematically summarized in Fig. 5.11.

Each station had a different period of record, so that the total number of storms available for analysis from KHA was 414 events; from KAS, 697 events; and from GED, 659 events.

Interstorm intervals for each of the three study sites are plotted in Fig. 5.11 as a set of scattergrams of the number of days to the next storm plotted as a function of the day of the year on which a storm event ends. This *time-to-next-storm* protocol causes a cluster of anomalously high values at the end of the monsoon season. The top panel in Fig. 5.11 shows that at Khartoum it is not at all unusual for the interannual dry period to exceed 200 days. However, in the case of Gedaref (the bottom panel), such a long dry period is exceptional. As one moves from the wetter south (bottom) to the drier north (top) in the figure, the plots show a systematic increase in the length of the interannual dry periods, as well as an end-

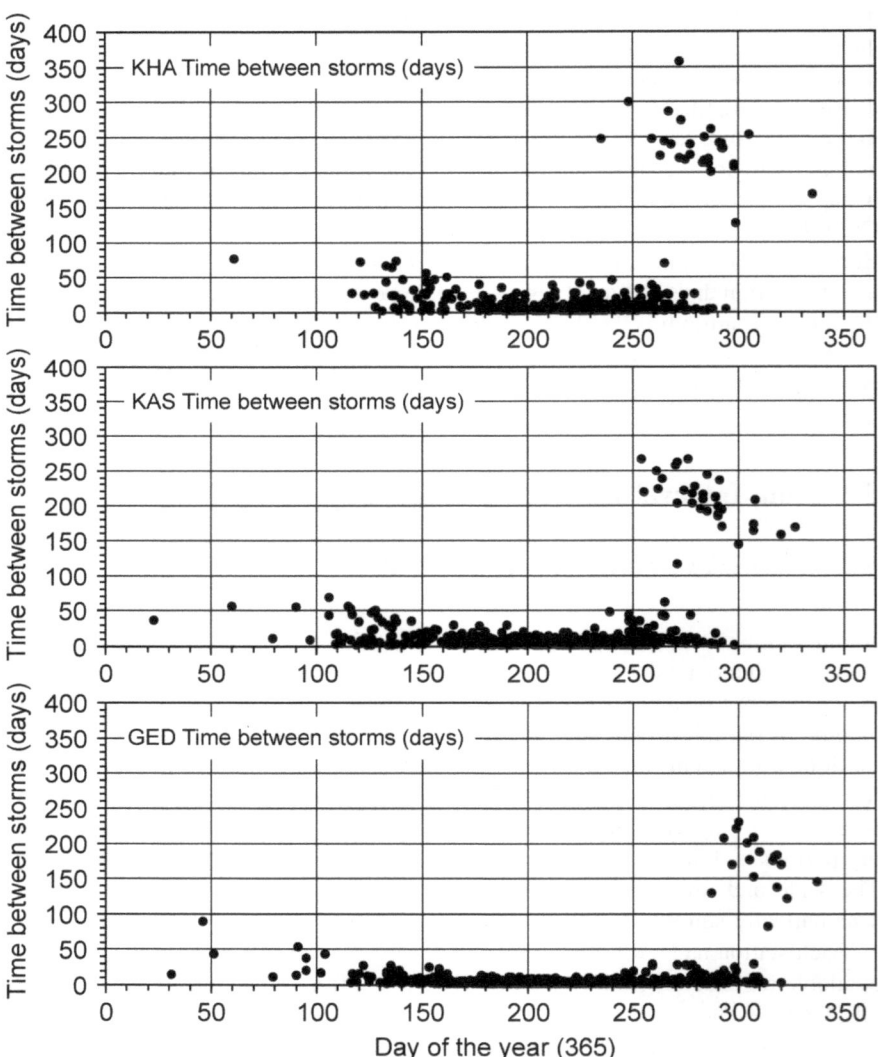

Fig 5.11 The time intervals between storm events recorded by the three stations: from north (*top*) to south (*bottom*), Khartoum (KHA), Kassala (KAS), Gedaref (GED)

of-season (EOS) broadening of the period over which large rain-free intervals might begin. Interstorm intervals are much shorter during the monsoon season for all three stations, with Gedaref having the longest, most persistent period of relatively frequent storms. Kassala is next, with Khartoum showing the shortest period of frequent storms, while having the longest intervals between storms during its rainy period and between its annual monsoon seasons.

5.2.2 Classification of Interstorm Intervals by Duration

The delay times in Fig. 5.11 are classified into binned intervals, and the number of events in each bin is normalized by the total number of events identified for the respective station. The normalized results (% within the specific bin) are plotted in Fig. 5.12 as histograms of the time intervals between storm events recorded by the three stations. The results are presented in terms of the class density of the probability (in fractional percentage) that the time interval between two events will fall within a particular bin. For example, if a storm occurs at KHA, there is a 13 % probability that the time to the next storm will fall in the range of 10–20 days. Depending on the crop and stage of growth, intervals longer than 10 days become a significant concern to agriculture.

Missing from the type of presentation in Fig. 5.12 is the time of year. Intuitively, one would anticipate that the expected time to the next event would be different following an event in November, with the looming expected interannual dry spell, than from after an event in July, say, in the middle of the rainy season. This is not accounted for by the plots in Fig. 5.12. Qualitatively, the portion of the histograms to the left in each panel—say, to the left of 100 days—generally represents the intraseasonal dry intervals within the monsoon season, whereas the small, secondary maximum to the extreme right (say, to the right of 100 days) represents the interannual dry intervals between monsoon seasons. Comparing Fig. 5.12 with Fig. 5.11, the three intervals from 10 to 60 days represent significant drought periods largely during the growing season, and collectively represent approximately 10 % of the interstorm intervals for Gedaref, 20 % of the interstorm intervals for Kassala, and 26 % of the interstorm intervals at Khartoum. The statistics summarized in Fig. 5.12 apply to the full 365-day calendar year.

5.2.3 Statistics for Dry Periods in the Wet (Monsoon) Season

However, if one is primarily interested in the statistics of intervals between rainfall for a typical growing season (the "wet" season) for agricultural planning, there is a danger that Fig. 5.12 might be mixing in events from earlier (or later) in the season. Figure 5.13 restricts the analysis to storm events during the growing (or monsoon) season.

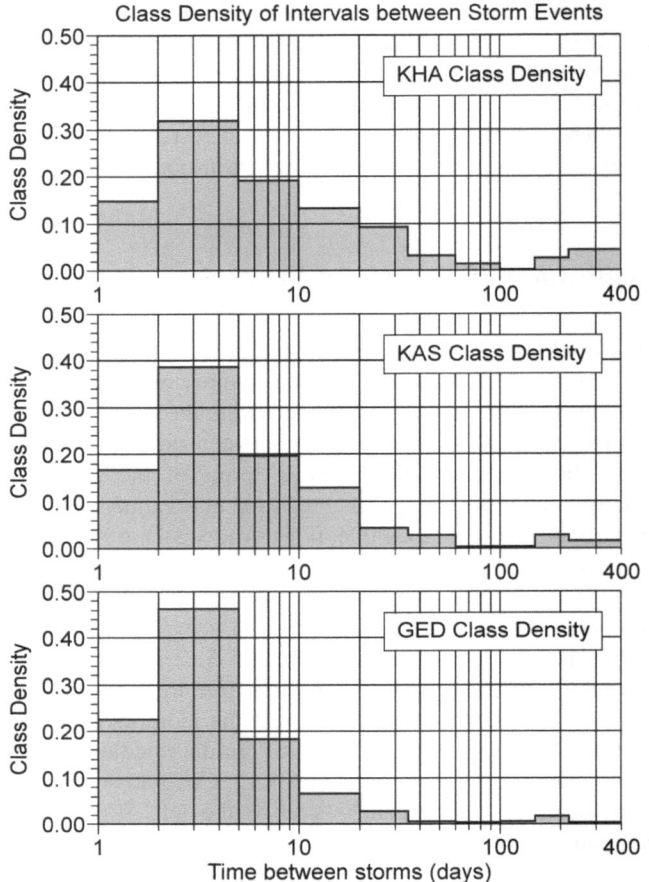

Fig. 5.12 Histograms of the time interval between storm events recorded by the three stations (from north (*top*) to south (*bottom*)): Khartoum (KHA), Kassala (KAS), Gedaref (GED). Bin edges are pseudo-geometrically spaced at 0, 2, 5, 10, 20, 35, 60. 100, 150, 220, and 400 days

First, a common calendar base is defined for the three sites over which the duration of the interseasonal dry periods will be characterized. Based on a visual inspection of the scatterplots in Fig. 5.11, the day on which a respective intra-seasonal dry period begins is restricted to an interval between calendar day 120 (30-Apr) thru calendar day 250 (7-Sep). To use events later in the season increases the likelihood of mixing interannual events with intraseasonal events, as should be apparent in the scatter plots in Fig. 5.11. For example, in the top panel, Khartoum has two extraordinarily long dry periods beginning in the range of calendar days before day 250: a dry period of 247 days, beginning on August 23, 1966 (calendar day 235); and a dry period of 300 days. beginning on September 5, 1970 (calendar day 248). Upon closer inspection of the multiannual time series, these two events

Fig. 5.13 Exceedence
probability for the number of
days between storms selected
for the intraseasonal rainy
period from Day 120 to 250

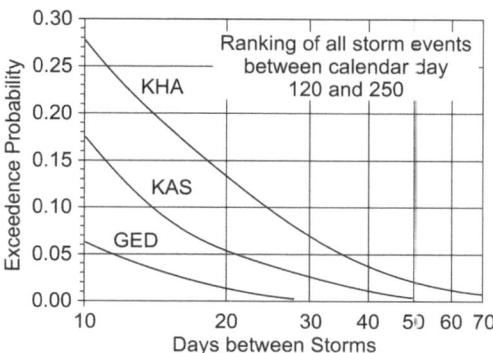

turned out to be interannual dry periods and were hence summarily dropped from
the analysis as being unrepresentative of intraseasonal dry-period events during the
wet period—the agricultural growing season—from Day 120 to Day 250. For this
analysis, the following numbers of events fell within the latter calendar interval:
472 events were available for GED, 566 events for KAS, and 319 events for KHA.

Figure 5.13 shows the probability (or frequency) that, if a dry period occurs
during the intraseasonal rainy period, its duration will exceed the number of days
along the abscissa. Since the focus here is on the frequency and duration of dry
periods impacting water availability, only the results for periods of 10 days and
longer are presented in Fig. 5.13. For Gedaref, dry interstorm intervals during the
wet period may exceed 10 days for 8 % of the events. For Kassala, dry intervals
may exceed 10 days for 19 % of the events; and for Khartoum 30 % of the dry
intervals may exceed 10 days.

For planning purposes, the results of Fig. 5.13 can be combined in different
ways with the previous analysis of storm events in this chapter, particularly with
the data in Table 5.1. For example, consider determining the return period of a
drought lasting 20 days or more at each station. Table 5.1 shows an expected 35
storm events per year for Gedaref, 21 storm events per year for Kassala, and 13
storm events per year for Khartoum. The probability that a dry period will exceed
20 days is obtained from Fig. 5.13, which shows that for GED it is 2 %, for KAS
6 %, and for KHA 14 %. In the case of Gedaref, 2 % of the interstorm dry periods
are longer than 20 days, and there is an expected number of 35 storm events per
year. Two percent of 35 events implies a recurrence frequency of 0.7 events per
year lasting 20 days or longer, or a return period of somewhat less than 2 years
(actually 1.42 years). For intraseasonal dry periods, these numbers lead to the
return period of a 20-day-or-longer drought being less than 2 years for GED and
less than 1 year for KAS; and for KHA, there is a frequency of recurrence or
expectation of approximately 2 such dry intervals during each monsoon rainy
season.

5.2.4 Statistics on the Interannual Dry Period

Perhaps the most predictable property of the Sahelian monsoons is that there will invariably be an interannual dry period from the end-of-season (EOS) last rain in the fall to the start-of-season (SOS) first rain in the spring or early summer (see, for example, Chap. 4, Fig. 4.5) What is not well predicted is the duration of this interannual dry period. To characterize the expected duration of this recurring annual drought and its variability from year to year, two parameters were tabulated for each year: the date of the end of the last storm event in each year (this defines the EOS for the year in question), and the time in days to the start of the first storm event in the following year (this defines the duration of the interannual dry period for that year). As shown in Fig. 5.11 and the daily-time series in Chap. 4, Fig. 4.5, GED has 18 EOSs and interannual dry periods in 19 years of record; KAS has 33 EOSs and interannual dry periods in 34 years of record; and KHA has 30 EOSs and interannual dry periods in 31 years of record. Bearing in mind that statistics from such small samples must be applied with caution, each parameter was then ranked by increasing value, and its cumulative Weibull statistic was computed. Results for both the timing of the EOS and the duration of the interannual dry period are summarized for each station in Table 5.6.

In addition, as a parallel, complementary computation, one might identify the "center of mass" (CM) or geometric center of each EOS point cluster by way of its horizontal and vertical positions in Fig. 5.11. These values are given in the first two rows of Table 5.6 for each station. The CM for GED occurs at day-of-the-year 309 (its x value), with an interannual interval (its y value) of 176 days until the spring storms might commence. For convenience, the coordinates of the CM are

Table 5.6 Summary of the durations of the inter seasonal hiatus in rainfall for the three reference stations (all units are days)

	GED	KAS	KHA
Mean EOS calendar day[a]	Day 309	Day 284	Day 279
Mean interannual dry period	176 d	210 d	237 d
Quantile stats on EOS and SOS			
Median EOS	Nov 5	Oct 10	Oct 11
Median SOS[b]	Apr 30	May 7	Jun 5
Q25(EOS)	Day 300	Day 271	Day 268
Q50(EOS)	Day 309	Day 283	Day 284
Q75(EOS)	Day 318	Day 292	Day 291
Stats on duration of interannual dry period			
Q25	143	186	217
Q50	176	209	237
Q75	203	232	249
10 % Exceed	221	260	287
Extreme	230	267	358

[a] Mean date of end-of-season (EOS; 365 day calendar) from CM
[b] Start of season (SOS) as median date of EOS plus median of dry period

abbreviated in the format (x_{CM}, y_{CM}), which in the case of GED becomes $(x_{CM}, y_{CM}) = (309, 176)$. The CM for KAS is (284, 210). The CM for KHA is (279, 237). In other words, as one progresses from south to north (see Fig. 5.11), the CM shifts to the left and upward. This implies that, as one moves north, the interannual hiatus in rainfall comes earlier in the fall season, from Day 309 (Nov 5) at GED, to Day 284 (Oct 11) at KAS, to Day 279 (Oct 6) at KHA, respectively. The shift upward is a consequence of the interannual dry period increasing in duration from south to north, which is enhanced by the onset of precipitation—the start-of-season (SOS)—the next year coming later in the spring as one moves northward. According to the CM analysis, the interannual dry interval increases from 176 days at GED, to 210 days at KAS, to 237 days at KHA, respectively.

The CM estimates in the first two rows of Table 5.6 are followed by the respective quantile statistics. For convenience, the expected end of season and start of season for each station are presented as dates. It is instructive to compare the mean EOS calendar day and the mean interannual dry period in rows 1 and 2 with the corresponding Q50 quartile or median statistic following in the table. Results agree quite well, with the KHA end-of-season results being the major difference. The CM EOS is estimated to be Day 279; however, the quantile analysis estimates EOS to be Q50 = Day 284, a 5-day difference, which might be expected from the different analytical methods used. The main points to take away from the quantile statistics in Table 5.6 are, first, the timing and variability of the end of season, and, second, the length and variability of the interannual dry period as summarized in Table 5.7.

Clearly, the timing of the start of season is more variable than the end of season at all three stations, with, interestingly, greater variability in the start of season at GED—a higher-rainfall site—than at KAS and KHA—lower-rainfall sites. Early-season, low-rainfall showers are more typical of Gedaref than of the inherently drier stations to the north. The interannual dry period is expected between the following dates at the respective stations: GED: Nov 5–Apr 30; KAS: Oct 10–May 7; KHA: Oct 11–Jun 5, with the variabilities in Table 5.7 estimated by dividing the IQR by 2.

The last two lines in Table 5.6 show that 10 % of the end-of-season interannual storm delays are greater than 221 days for GED, greater than 260 days for KAS, and greater than 287 days for KHA. The extreme of all values for the entire dataset is 358 days for Khartoum. This event is associated with the well-documented 1984 drought in the northern Sahel. At Khartoum it covers the period from September

Table 5.7 Compare variabilities of EOS, SOS, and interannual dry period

Parameter	GED (d)	KAS (d)	KHA (d)
EOS (Day of year)	Day 309 ± 9	Day 283 ± 11	Day 285 ± 12
Interannual dry period	176 ± 30	209 ± 23	237 ± 15
SOS (Day of year)	Day 120 ± 31	Day 127 ± 25	Day 157 ± 19

These values are based on quintiles of ranked samples. The estimated ranges are based in 1/2 the interquartile range (or IQR/2)

29, 1983, following a fairly typical, but reduced, rainy season in 1983 of 83 mm rainfall, to the first rain in the following year on September 22, 1984, a duration of 358 days. However, the first event in 1984, on September 22, measured only 3.3 mm, and in any practical sense the drought continued until the minor storm on May 21, 1985, having a rainfall total of 7 mm. Therefore, discounting several minor events of less than 1 mm rainfall, in 1984–1985, Khartoum experienced 600 days with a total accumulation of less than 5 mm.

References

Balme M, Vischel T, Lebel T, Peugeot C, Galle S (2006) Sahelian water balance: impact of the mesoscale rainfall variability on runoff. Part 1: rainfall variability analysis. J Hydrol 23:336–348

Bell MA, Lamb PJ (2006) Integration of weather system variability to multidecadal regional climate change: The West African Sudan-Sahel zone, 1951–98. J Climate 19(20):5343–5365

D'Amato N, Lebel T (1998) On the characteristics of the rainfall events in the Sahel with a view to the analysis of climatic variability. Int J Clim 18:955–974

El Tom MA (1975) The rains of the Sudan: mechanisms and distribution. Khartoum University Press, Khartoum

GHCN-Daily (2012) The global historical climatology network, operated by the NOAA national climate data center, provides historical daily data exchanged under the world meteorological organization (WMO) world weather watch program. http://www.ncdc.noaa.gov/oa/climate/ghcn-daily/. Accessed 20 Oct 2012

Guy N, Rutledge SA (2012) Regional comparison of West African convective characteristics: a TRMM-based climatology. Q J R Meteorol Soc 138:1179–1195

Hulme M, Trilsbach A (1989) The August 1988 storm over Khartoum: its climatology and impact. Weather 44(2):82–90

Lebel T, Ali A (2009) Recent trends in the central and Western Sahel rainfall regime (1990–2007). J Hydro 375:52–64

Nicholson SE (2009) A revised picture of the structure of the "monsoon" and land ITCZ over West Africa. Clim Dyn 32(7–8):1155–1171

Sutcliffe JV, Dugdale G, Milford JR (1989) The Sudan floods of 1988. Tech. Note, Hydrol Sci 34(3):355–364

Chapter 6
Overview and Conclusions

Abstract Sustainable cultures and economies in the East Sahel need to evolve with a strong base at the local community level. In a region where water is such a determining asset, this requires a fundamental adaptation to the constraints that nature itself places on the resource. The greatest, most fundamental, and most immediate threat to the local population is the unpredictable variability of rainfall—the dimensions of which can only be appreciated by understanding the historical record on monthly, annual, and multidecadal time scales. However, due to the specific nature of the dominant type of storm systems in the region, certain essential aspects of rain storm patterns cannot be captured without daily coverage. While most of the rain days have low intensity, most of the annual rainfall total is contributed by the fewer, high-intensity events associated with deep convection into the troposphere. Although the most extreme storms often leave little to no imprint on monthly or annual totals, these are the events that most significantly impact the local landscape and communities. An overarching concern by many is a pessimistic view that climate—namely rainfall—in the East Sahel has degraded significantly over the last decades and is expected to continue to do so into the coming century. In fact, there is little evidence to support such claims. Climate change is occurring in the East Sahel, but is largely dominated by relatively incoherent fluctuations, not regionally coherent monotonic trends of significant magnitude. Mitigating the risks from the type of extreme climate variability that is evident in the historic rainfall record underscores the imperative to adapt water policies to the local scale. If local communities can be empowered to deal with the type of present-day natural behaviors described here, they should be well-prepared to deal with the pending uncertainties of the future.

6.1 General Patterns of Spatial Variability

It was noted in Chap. 1 that the international community is coming to realize the importance of placing the management of local development projects in the hands of local communities, transitioning from broad regional programs to local, indigenous, self-sustaining projects. In the Sahel, south of the Sahara, the management of water is one of the most pressing issues—the primary source of water being the annual monsoon cycle. It would seem wise, therefore, as future plans are formulated, to refocus water policies through the lens of local climate, its trends, and its variabilities. Since all engineering projects are ultimately constrained by the natural limits imposed by nature, it follows that water projects—whether a major dam on the Atbara River, or the diversion of hill-slope runoff into a rain-fed agricultural plot in Kassala—are clearly bound by the availability of rainfall. This report draws on historical rainfall data from a local corner of the East Sahel to address the following questions:

(1) Where does it rain?
(2) When does it rain?
(3) And how much does it rain?

   Simple answers—such as climate normals—are not acceptable for the community addressed here. The consequences of each response demand that it characterize the intrinsic variability of the answer in space and time. Having outlined earlier in the text a number of complementary methodologies for describing relatively local, short-term, intraseasonal rainfall variability in the context of long-term, regional hydroclimate patterns, in this, the final chapter, the major objectives and selected findings are reviewed, particularly emphasizing those aspects of the hydro climate relevant to local water management. The topics are chosen to underscore the thesis of this report—that, for purposes of coping with climatic factors in the immediate future, it is the intraseasonal and interannual variability of rainfall that is the principal threat to the well-being of local communities in the East Sahel—with, perhaps, the results from the East Sahel serving as a proxy to the larger Sahel in general.

6.1.1 Spatial Gradient of Annual Rainfall

The Sahel—defined here as the area south of the Sahara between the 100 mm/year isohyet to the north and the 600 mm/year isohyet to the south—is only 250 km or so wide in the east of Africa (Fig. 6.1). A strong north-to-south spatial gradient of annual rainfall in the East Sahel of some 2.2 mm/km is quite evident in the maps of annual precipitation (Figs. 1.1, 1.3, 1.5); also in the seasonal histograms of expected monthly precipitation for the three reference sites (Fig. 1.6); in the time series of annual totals in (Fig. 2.2); and in the multi-year time series of monthly values (Figs. 3.1, 3.2). While this north–south spatial pattern is similar to that in the West Sahel, where Balme et al. (2006) report a north-to-south gradient of approximately 1 mm/km, the gradient in the East Sahel appears to be somewhat more intense. One reason for this difference is that rainfall patterns in the East Sahel are affected by orographic effects from the Ethiopian Highlands to the south and southeast (Figs. 1.5; 6.1).

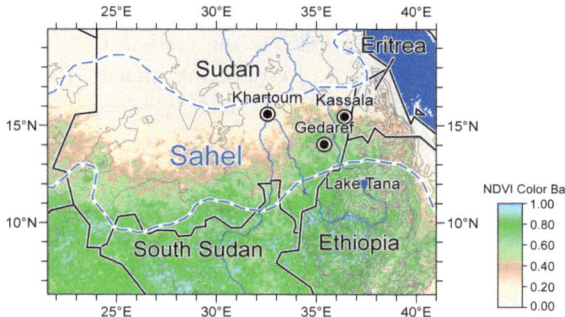

Fig. 6.1 The East Sahel showing the three WMO GHCN station gauges used for this study with a map of the Normalized Differential Vegetation Index (NDVI) from the Africa GIMMS 8 km database[1] composited for August 15–30, 1988, following the storm of record (August 4–5) at Khartoum. *Blue dashed lines* are the 100 mm (north) and 600 mm (south) isohyets of annual rainfall used to define the bounds of the Sahel. Elevation contours are at 500 m intervals. The *color bar* for the NDVI is shown

6.1.2 Interannual Variability of Annual Totals

In addition to the broad spatial variability of annual totals among the three reference stations used in this study, there is a strong interannual variability of annual totals at each of the stations as defined by the interquartile dispersion (IQD; see Sect. 2.4). In the East Sahel, there is a systematic pattern in the manner of how this interannual temporal variability varies from station to station consistent with other studies. Jurkovica and Pasaric (2012), for example, analyzing 50 years of ground station precipitation data (1951–2000), show that, on a global scale—including marine areas—the precipitation in the rainbelt associated with the Intertropical Convergence Zone (ITCZ) tends to have less relative variability than typical rainfall elsewhere on the globe. They argue, however, that the African Sahel is an exception. The northern Sahel in their study shows higher-than-normal rainfall variability (relative to local norms), and the southern Sahel shows less-than-normal variability, with the demarcation line being approximately the 500 mm isohyet. This implies an inverse relationship between relative variability and annual totals—the higher the annual total for an area, the lower its relative variability. This pattern accords with the results of others, and with this report. See, for example, the interannual variability of annual totals represented by the interquartile dispersion (IQD) coefficients

---

[1] Background map is composed from gridded (8 km) values of the normalized difference vegetation index (NDVI) derived for Africa from the Advanced Very High Resolution Radiometer (AVHRR) instrument onboard the NOAA-9 satellite (Tucker et al. 2005). Values of 0 correspond to no vegetation (bare soil). Values of 0.7 and higher denote dense vegetation Values of 0.2 to 0.4 are typical for open savannas and grazing land. Higher than normal rainfall was experienced in the northeast Sahel beginning in late July, with the major storm at Khartoum on August 4–5, 1988, when rains at KHA were significantly heavier in actual terms than at KAS or GED (Figs. 2.2, 3.2.). Original NDVI data are available at Africa GIMMS NDVI (2013).

in Tables 2.3, 2.4 and 2.5, and Figs. 2.4, 2.5 and 2.6. The IQD is defined as the ratio of the interquartile range, IQR = Q75–Q25, to the median Q50. Thus IQD = IQR/Q50. Rounding off the IQD values to two decimal places, for the southernmost (wettest) station: Gedaref, IQD(GED) = 0.26 (P^{Total}_{annual} = 600 mm); for Kassala, IQD(KAS) = 0.44 (P^{Total}_{annual} =280 mm); and finally, for the northernmost (driest) station, Khartoum, IQD(KHA) = 0.62 (P^{Total}_{annual} = 130 mm). This inverse correlation between interannual variability and annual totals also appears in the interannual variability of monthly totals for the East Sahel (Tables 3.1, 3.2 and 3.3, Fig. 3.3), where, for example, for the principal month of August, the IQDs for the respective stations are: IQD(GED) = 0.56; IQD(KAS) = 0.74; and IQD(KHA) = 1.22.

It is important to keep in mind, however, that these measures of variability—in fact most metrics for variability used in the literature—are relative to some type of local norm; for example the IQD is a metric that the IQR is normalized by the long-term median value, Q50, of the respective parameter. The long-term (100+ year) spatial trend of the un-normalized IQR of annual rainfall totals is in the opposite sense than the IQD geographically: IQR(GED) = 154 mm; IQR (KAS) = 122 mm; and IQR(KHA) = 81 mm (Table 2.3). Thus while the IQD *increases* from wet conditions in the south to dry conditions in the north, the IQR *decreases* from south to north. In some applications, the actual or absolute variability might be of more consequence than the relative or normalized variability.

6.2 Multi-Station Versus Single-Station Metrics

6.2.1 Un-normalized Versus Normalized Metrics

The direct comparison of individual time series of annual totals from the three study sites qualitatively illustrates the magnitude of interannual variability (Fig. 2.2). However, to more closely compare the relative behaviors among stations, it may be necessary to *normalize* the respective time series in some way, particularly if the objective is to combine a number of individual station time series having different statistics (e.g. different mean amplitudes, or different characteristic variabilities) into a single multi-station composite times series. Thus, it is common practice (cf. Chap. 2) to reduce the annual totals from each station to a time series of normalized metrics such as

Type 1: standardized precipitation indices (SPI; Fig. 2.8);
Type 2: each year's total as a percent of the expected annual value (Fig. 2.10);
Type 3: each year's total as a percent departure from the expected annual value (Fig. 2.11).

Note that the Type 3 metric might be thought of as the time series equivalent of the IQD. Each of these three types of procedures offers a somewhat different perspective for investigating similarities and differences in the relative morphology—or *shape*—of interannual behaviors among respective sites, regardless of the actual expected magnitude of the annual rainfall at each site. In addition, such normalized metrics can be composited from a number of sites such that the data from sites having the highest expected annual total or greatest expected variability do not overwhelm the contribution from sites having lesser totals. In other words, the composite time series of such normalized values allows one to better identify common patterns of temporal variability among sites.

Although each normalized metric is widely used in the climate and climate modeling literature—each offering its own advantages for summarizing features of the data—perhaps the most common of these procedures is the SPI or standardized precipitation index (Fig. 2.8). As explained in Chap. 2, standardization is a common statistical operation that, in the case of the precipitation index, involves subtracting the expected long-term mean value for a station from the individual annual total for each year over the period of record, then normalizing by the standard deviation from the expected mean for the station. When compositing data from multiple stations, Kraus (1977) suggests that the SPI offsets some of the bias associated with the contribution from stations typically having higher variability. Nicholson (1986) asserts that converting annual rainfall totals to standardized values (the SPI) prior to compositing enables the merging of mixed datasets having diverse statistics, with the advantage of "smoothing" over local interannual variability between stations. In other words, the composite SPI smooths out the "noise" from extreme weather events recorded by only one or two of the stations in a large multi-station array. For example, Nicholson contends—based on intercomparing sets of regionalized multi-station composites of SPIs—that "on an annual time scale, rainfall tends to exhibit coherent patterns of variability over most of the African continent," with concomitant implications for possible large-scale spatial and temporal teleconnections between African climate and fundamental processes in the global oceans and atmosphere. Ali and Lebel (2008) also acknowledge the value of a Sahel-wide, multi-station composite SPI as a relatively robust indicator of regional conditions, so that it is logical that climate modelers (e.g. Giannini et al. 2003) have found regional multi-station SPI composites (see, for example, Figs. 2.7 and 2.8) to be essential for understanding the primary factors modulating the climate of the Trans-African Sahel (defined by some for the purpose of climate modeling as the region: 10°N and 20°N, 20°W and 40°E).

Alternatively, some workers prefer other normalized composites. Held et al. (2005) developed a rainfall composite by normalizing surface gauge data by their long-term mean value—the Type 2 procedure in the list above. (Additional smoothing in their case was ensured by applying a 5-year running average to the annual composited data.) By combining regionalized composite rainfall metrics with global climate models, it is now widely accepted that—on the spatial scale of the Trans-African Sahel—the principal source of interannual to interdecadal rainfall variability is from fluctuations in sea surface temperatures (Giannini et al. 2008).

However, the spatial scale of these regional climate reconstructions is in marked contrast to the single-station analyses emphasized in this report (Chap. 2), which would seem to suggest that such spatial coherency is often the exception, rather than the rule. This latter view accords with observations in the West Sahel (D'Amato and Lebel 1998; Agnew 2000; Balme et al. 2006; Lebel and Ali 2009). Multi-station composites, such as the SPI (Fig. 2.8), often mask interannual variability of local rainfall among the stations themselves, as pointed out a decade ago by Agnew (2000) who used 35 stations in the continental (i.e., non-coastal) West Sahel to compute single- and multi-station time series of standardized precipitation anomalies. Accounting for the different time base he used, his composite SPI time series—while not reproduced in this report—is quite similar to others published for the West Sahel (e.g. Figs. 2.7 and 2.8). In addition, Agnew superimposed the maximum and minimum annual anomalies from individual stations, which, he points out, show significant negative and positive annual excursions of local areas from his regional, 35-station composite. He suggests that care be taken when literally interpreting too much detail from the spatially averaged, composite SPI, and advocates for analyzing the individual SPI anomalies for each station, rather than relying solely on spatially aggregated anomalies. Agnew is not alone in this opinion. Ali and Lebel (2008), while espousing the value of composite SPIs, also caution that the regional composite masks the strong underlying spatial variability of rainfall. Bell and Lamb (2006) assert that "such a distinctly 'climate' perspective (the composite SPI) overlooks the weather system variability that produces the seasonal or annual rainfall departures featured in such time series."

With regard to the East Sahel, and the three-station composite SPI presented in Chap. 2 (Fig. 2.9), over the last several decades (say from 1990 to 2009) the aggregated time series appears to be somewhat depressed (-0.2 sd) from the long-term expected (median) value, similar to the W. Sahel JISAO SPI of -0.3 sd for its last 10-year period of record (1995–2004). However, what is not so evident in the three-station composite is the Trans-African desiccation of the 1970s through the 1980s, that is so well developed in the W. Sahel JISAO SPI in Fig. 2.9. Inter-comparing the individual station SPIs in the East Sahel in this figure, it appears that 1984 and 1990–1991 were extreme drought years in-common among the three stations, but the longer term Sahelian "desiccation" is actually interspersed with "wet" years in the East Sahel. As for long-term trends in common for these three stations, over the last decade and a half, the SPI for KHA is approximately -0.3 sd, and for KAS about -0.5 sd; thus both are somewhat depressed from the long term expectation. However, the SPI values for GED are closer to $+0.1$ sd, hence elevated, suggesting slightly wetter conditions than the long-term conditions would imply from the century-long period of record. Year-by-year, and multi-year conditions might be quite dissimilar among these three stations with concomitant implications for the local variability of climate.

Thus, while there is little doubt regarding the efficacy of multi-station composite metrics to suppress interstation spatial variability for large-scale regional and sub-continental studies, the consequence is that from the point of view of local climate variability, what is *noise* to some might be *signal* to others. This report sees the interstation variability as the *signal* of the true nature of the Sahel climate (see also Hulme 2001). And the characterization of this signal—which is to say that knowing the statistics of interseasonal and interannual anomalies on the local scale— is essential to water management, from predicting the likelihood of, and responding to, flash floods, to anticipating and managing intraseasonal and inter-seasonal droughts. The vast percentage of agriculture in the East Sahel is rainfed, much of it mechanized, except for small-scale family gardens. But also significant are the uncultivated land areas in the region having rainfalls of 200 mm or less; these are used for livestock grazing—cattle, camels, sheep, and goats—of significant economic value to the local economy in their own right. A few weeks' hiatus in seasonal rainfall might be a "blip" on the screen of international NGOs, but could be devastating to a local community of pastoralists or nomads.

6.2.2 Interannual Patterns of Rainfall Variability for the East Sahel

6.2.2.1 Spatial and Temporal Interannual Variabilities

It is noteworthy (Fig. 2.10) that annual rainfall in Khartoum fluctuates from, occasionally, 250 percent of the normal value to, often, 50 % of the normal value, and occasionally to negligible or no rainfall for the entire year. Regional, sub-continental scale, multi-station composites for the Trans-African Sahel over this same time period are significantly more subdued, fluctuating from approximately 125 % of the expected annual rainfall to 75 % of the annual rainfall (Nicholson 1989; Held et al. 2005). Typically, the relative interannual variability is twice as large or larger in the northern Sahel compared to the southern Sahel, whether using regional composites (Fig. 2.11) or single-station metrics (Fig. 2.13; note the difference in scale between KHA and GED). This is also shown by the interquartile dispersion (IQD) values in Tables 2.3, 2.4 and 2.5. For the northernmost station Khartoum IQD(KHA) = 0.62, whereas for the southernmost station, Gedaref, IQD(GED) = 0.26. The extreme character of interstation variability for annual "events" is also apparent in the normalized time series (Fig. 2.9), for example, where the annual difference in the SPI at two stations in the East Sahel can easily exceed 2 sd. Multi-year "clusters" of the SPI having similar sign (wet or dry) are typically of 3–5 years in duration, with interstation annual anomalies being greater than one sd.

The time series of the percent departure from the expected annual total for each station $((\Delta P(\%)_{ij}$, where i is the year, and j is the station; see Fig. 2.13) provides a somewhat different perspective when comparing the amplitude of the relative variability of a station, with its amplitude at another station. Compare, for example, the "wet" decades from 1950 to 1970 at the northernmost station, Khartoum, with Gedaref, the southernmost station, for the same period (Fig. 2.13). Rainfall at Khartoum is typically 50 percent above normal, whereas rainfall at Gedaref fluctuates ±15 % or so above normal. The composite sub-continental trans-African metrics of Nicholson (1989) (Fig. 2.11) imply a significant tilt toward the behavior of rainfall variability at Khartoum (Fig. 2.13), although more subdued with a positive bias of only approximately 15–20 % above normal, rather than the 50 % or more at Khartoum.

Note that the 1960 year-long loss of rainfall in the northern group of stations of Nicholson (see Fig. 2.11), while not so evident in her southern group of stations, does have its counterpart, albeit variable, among the three stations' composite in the East Sahel (Fig. 2.13).

The implication of these results for the East Sahel is that for disaster mitigation, as well as for properly managing water for domestic and agricultural uses, understanding the interannual variability of water supply to a community is as important as—or, arguably, more important than—knowing the long-term expected annual rainfall for a region. This is because, over multiple generations, a local culture or community will adapt—by choice of crops and herding techniques—to the long-term expected rainfall in a region, whether the annual total is 150 mm or 600 mm. From the point of view of climate, it is the *unpredictability* of the interannual variability of this rainfall that poses the greatest threat to the population.

6.2.3 Is Interannual Variability Increasing in Time?

Chapter 2 shows that interannual variability is a common feature of precipitation records from the East Sahel over the last century. Ali and Lebel (2008), using multi-station time series from 1950 to 2006, suggest that, in fact, the Sahel may have experienced an *increase* in interannual variability over the last few years. However, this is not readily apparent for the East Sahel from the evidence in this report, namely the 100+ years of annual totals (Fig. 2.2), as well as the various time series of normalized metrics in Sect. 2.5 (Figs. 2.9, 2.10 and 2.13). In the East Sahel, the magnitude of the interannual variability seems to be masking local evidence of any systematic change in the scale of the variability. According to the data developed in Sect. 2.5, the period of 1920 to 1940 is quite unsettled (i.e., *variable*) for all three of the stations in the East Sahel, with the period from 1920 to 1970 being most unsettled (i.e., having the greatest departures from the expected annual total) for the northern portion of the East Sahel, namely the region about Khartoum. Indeed, with reference to the percent departures from annual total rainfall in Fig. 2.13, the interannual variability at either Khartoum or Gedaref, over the last 15 years, seems to be no more extreme in these time series than during other epochs during the respective 100+ year period of record.

6.3 Interannual and Interseasonal Variations in Monthly Rainfall

Multiannual time series of monthly rainfall from the three study sites (Chap. 3) show a summer-long, single-mode, annual monsoon cycle, with virtually no rainfall from the mid-fall to late spring (Fig. 3.1). Precipitation can be episodic over the course of the rainy season, with significant year-to-year or interannual fluctuations in annual precipitation totals, as well as the intraseasonal timing of the month of maximum rainfall—which shifts irregularly from year to year between August and July, and occasionally beyond. Moreover, although the expected annual rainfall for Khartoum (130 mm) is significantly less than that of Kassala (280 mm) or Gedaref (600 mm), monthly rainfall at Khartoum occasionally exceeds that at Kassala, and occasionally rivals that at Gedaref (Figs. 3.1, 3.2).

The interannual variability of monthly rainfall over the 100+ years of record for each station often shows a pattern of multi-decadal modulations of monthly totals (Fig. 3.2), some of which are identifiable in the annual data (Fig. 2.2). According to the literature (such as Nicholson 2011, Giannini et al. 2008), many of these long-term modulations are explained by the global teleconnections in the atmosphere between local climate and long-term fluctuations of sea surface temperature (SST). However, the type of differences among local stations that are apparent in the monthly time series (Figs. 3.1, 3.2) and the annual time series (Fig. 2.2) are more difficult to understand in terms of these teleconnections, and pose significant challenges to climate modelers (Biasutti et al. 2008). A gentle, multi-decadal modulation of monthly rainfall at Gedaref (Fig. 3.2), having a characteristic period of 13 to 15 years, is asynchronous with the two other reference stations (KAS, KHA) in the East Sahel, suggesting the influence of more local factors. Factors on a different time scale are responsible for the examples of interseasonal and year-by-year interannual variability between stations (Fig. 3.1) causing the spatial variability in the intensity of local floods (e.g. GED: 1982; KHA: 1988; KAS: 1974) and droughts (e.g. KHA: 1984; KAS: 1966). Much of this spatial and temporal detail is lost when data from many stations over large areas of the Sahel are aggregated into regional normalized precipitation indices as in the case of SPI time series (Figs. 2.7, 2.8), or percent departure from the annual mean (Fig. 2.13).

Classifying the 100+ years of monthly data into the expected totals for each of the 12 months at the three reference stations (Fig. 3.3, Tables 3.1, 3.2 and 3.3) shows that the monsoon season becomes longer as one moves from station-to-station, north to south, and the expected monthly total rainfall increases at each respective site (Fig. 1.3). The graphed and tabulated quantile analysis of the binned monthly values (Fig. 3.3, Tables 3.1, 3.2 and 3.3) indicate that 93 % of the annual precipitation for Gedaref occurs from June through September (JJAS), with the interquartile dispersion (IQD) varying inversely with the amount of monthly precipitation—months of higher precipitation totals tend to have lower interquartile dispersion. For Kassala, 96 % of the annual precipitation occurs from June through September (JJAS), with monthly interquartile dispersions being measurably greater than those for GED. The monsoon season at Khartoum is significantly shorter, with 96 % of the annual precipitation occurring in the three months from July through September (JAS), with the interquartile dispersion for these principal monsoon months substantially greater than for the other two stations, GED and KAS.

The three-station study area in this report is in one of the regions in East Africa identified as EA1 (or *East Africa 1*) by Nicholson and Selato (2000), who divided

Africa into regions having similar rainfall patterns. However, the composite of the seasonal distribution of precipitation for the three-station study area in the East Sahel is significantly more compact, seasonally, than the regional EA1 distribution (Fig. 3.4), which is not surprising since some of the stations in the EA1 area of Nicholson and Selato were further to the south than the current study area, thus recording a contribution from lower latitude, more humid stations, and swept by the annual latitudinal march of the ITCZ tropical rainbelt for a longer period of the season. This points out, however, the important role that spatial variability plays in the temporal variability of the delivery of water to the East Sahel, and the necessity of adjusting one's database to the precise geographic region of interest. This would apply to climate modelers as well as to water managers.

6.4 Months of Maximum Precipitation and Greatest Variability

There are aspects of the intraseasonal variability in patterns of rainfall that are critical to rain-fed agriculture and local—even regional—water management. For example, certain studies in the West Sahel assert that "August is the wettest month," and the pattern of rainfall in August "contributes disproportionately to the interannual variability." However, it may be misleading to invoke these general-izations as a rule-of-thumb for the entire Sahel (Chap. 3). For example, in terms of the seasonal timing for maximum precipitation in the East Sahel, the probability that the annual maximum monthly rainfall will occur in July is two-thirds that of occurring in August, the generally held view (Fig. 3.5, Table 3.4). In other words, from the viewpoint of a water manager, or a community depending on rainfed agriculture, for every 10 years that the seasonal rainfall maximum occurs in August, there will be almost seven years for which the seasonal rainfall is maxi-mum in July. Of course, there is also a 10 % probability (Table 3.4) that the seasonal maximum will not occur in either month, but the main point here is to emphasize that the frequency of occurrence of monthly maxima in July is par-ticularly significant.

As to the question of which time of the season is most responsible for inter-annual variability of annual totals, monthly data for three months (JAS) are compared from the three reference stations. The monthly relative variance (RV) is defined as the ratio of the interquartile variance of a month's totals to the sum of the interquartile variances for all three months (JAS) in the season (Table 3.5). Hence, the RV is a measure of the expected partitioning of the variability of rainfall for a specific month relative to the overall variability of the annual total at a station. The RVs for GED (Table 3.5)—the "wettest" station—indicate that rainfall in August is responsible for 60 % of the interannual variance in annual totals. For stations of lower expected annual rainfall, July becomes a major con-tributor to the interannual variability, so that for the 105 years of available data from Khartoum, the interannual variability in the annual rainfall total is driven as much by variability of the monsoons in July as those in August (Table 3.5).

Therefore, for some applications, analysts might need to adjust the commonly held notion that August is the wettest month and contributes disproportionately to interannual variability of rainfall in the Sahel. In addition to climate science, there is a community of users—particularly those stakeholders concerned with the local availability of water—that might be concerned with a more robust picture of the interannual timing of monthly maxima and variability. If so, such quantitative investigations in the East Sahel need to include at least July and August, and even these 2 months alone may not be sufficient for many applications. Climate modelers often use 3 months (JAS) to simulate total annual rainfall for the Trans-African Sahel (cf. Held et al. 2005; Giannini et al. 2008). However, even these 3 months (JAS) are not truly representative of seasonal totals throughout all of the Sahel, particularly for areas where the annual totals exceed several hundred mm. Stations in such areas tend to have longer rainfall seasons than the JAS months. In the case of Gedaref, for example, approximately 20 % of the annual rainfall of 600 mm occurs outside of the JAS season (see Table 3.1). Thus climate studies of the Sahel employing precipitation totals from JAS might systematically be underestimating the potential contribution of rainfall from the southern Sahel to regional composites. This may not be inconsequential for models purporting to address issues relevant to indigenous cultures and sustainable economies.

6.5   Statistics of Storms: Duration, Frequency, and Total Rainfall

6.5.1   Importance of Daily Rainfall Data for Water Management

Major storm and consequent flood events of 1964, 1974, and 1988, along with the regional drought events of 1966, 1984, and 1985, are well documented in both the monthly data (Figs. 1.4, 3.1), and the daily data (Fig. 4.4) from the three study sites. However, it is important to also recognize that, upon close inspection of these data, there are some extreme daily storm events (Table 4.1) exceeding 80 mm that do not have counterparts in anomalous annual totals, or even in anomalous monthly values. For example, consider the extreme daily event of 172 mm at Gedaref in 1958 (Fig. 4.4), or the 100 mm daily event in 1970, neither of which has significant counterparts in the annual data from Gedaref (Figs. 2.2, 4.4), or even in its monthly values (Figs. 3.1, 3.2). Also consider the extreme 103 mm daily event in 1983 at Kassala that has no anomaly in the annual value for the station (Figs. 2.2, 4.4), and has no anomalous value in its monthly time series (Figs. 1.4 or 3.2).

Unfortunately, most historic climate studies rely on monthly and annual time series, and it is common for many areas of the globe that daily data are simply not available for long periods of record through organizations like the WMO. Thus, many of the details on the patterns of storm events in the past are not available in the public record. Losing this information profoundly undermines the development of the tools needed for managing the impact of flash flooding—and intraseasonal dry periods—on ever-increasing populations. Because of the limited quantity of daily data that are available, this report at best provides a rudimentary insight into the nature of short-term rainfall (and droughts) in the East Sahel.

6.5.2 Spatial and Temporal Character of Storm Days

If one defines a "rain day" (or "storm day") as a 24-h recording period having a rainfall total of 1 mm or greater, then the number of rain days per year for each of the three study sites are GED: 56 days; KAS: 30 days; KHA: 15 days. This pattern—the number of rain days per year—is roughly apparent in the respective daily time series (Figs. 4.3, 4.5). The daily data in the East Sahel show that while most of the rain days have low intensity (mm/day), most of the annual rainfall is contributed by the fewer, high-intensity events (Fig. 4.6), those which are likely associated with intense, local-scale, convective-storm systems. The daily rainfall data show that the monsoon season in the Northeast Sahel (KHA) tends to be 20–30 days shorter than the rainy season to the south at KAS and GED. The midseasonal behaviors of the three datasets are quite similar (Fig. 4.8), with close-to-identical values for the cumulative percent (60 %) of annual precipitation by Day 225 (Aug. 13), and 50 % of the annual rainfall occurring within a day or two of Day 217 (Aug. 5). Analytical "models" derived in this report representing the expected seasonal distribution of daily rainfall indicate that the expected timing of peak daily precipitation for the year, in percent of the annual total, typically occurs about Day 223 (Aug. 11), differing among these three stations by only a matter of a few days. The statistical expectation, however, significantly under-represents the effect of interannual variability of daily rainfall patterns.

6.5.3 Duration and Frequency of Storm Events

A storm *event* is defined here as a specific episode of recorded (i.e. *measurable*) precipitation having a continuous duration of one or more days, and a total accumulation equal to or exceeding 1 mm (see Chap. 5). As a generalization, it is unusual for storm activity to continue over many contiguous days in the East Sahel. For Gedaref, 86 % all storms have durations of three days or less (Tables 5.1a–c), and no storm has been recorded lasting more than 6 days over the 19-year period of record from Gedaref, or the 34-year period of record from Kassala, or the

31-year period of record from Khartoum. For Khartoum, in fact, there were no storms of duration longer than 4 days. The majority of storm events at all three stations (Tables 5.1a–c) have durations of no longer than one day. One-day storms account for 69 % of all events at GED; for 80 % of all events at KAS; and for 82 % of all events at KHA—clearly a consequence of the type of localized, shor-term, spatially isolated convective storm systems responsible for the majority of rainfall events in the Sahel (El Tom 1975, El Gamri et al. 2009).

Comparing the magnitude of storm events by their number per year versus their respective contribution to the overall annual total, the class of events having totals less than 30 mm, for example, accounts for 80 % of the total number of annual events at Gedaref (Fig. 5.2)—most events have low rainfall. However, events having totals less than 30 mm contribute only 44 % of the annual rainfall total at Gedaref (Fig. 5.6). Most of the annual rainfall (56 %) comes from events having storm totals greater than 30 mm. Thus, while there are many more smaller events than larger events, it is the fewer, larger storms that deliver a proportionately greater percentage of the annual rainfall, a pattern that applies to the storm events at all three reference stations.

Next, there is the question of the relative magnitude versus number of storm events contributing to the gross differences in the expected annual rainfall between each station (Table 1.1). Are these differences largely due to differences in the magnitude of discrete events, or to differences in the number of events at each site? The total annual rainfall at Kassala (280 mm) is approximately half that at Gedaref (600 mm), and the total annual rainfall at Khartoum (130 mm) is approximately half that at Kassala, or a quarter that at Gedaref. In comparison, expected storm event magnitudes at Kassala (7 mm) and Khartoum (7 mm), presented by the median storm totals (Tables 5.1a–c) are approximately 64 % smaller than those at Gedaref (11 mm). In addition, expected storm event magnitudes at Kassala and Khartoum contributing to their respective annual totals are only 67–76 % of those at Gedaref (Table 5.5). The greatest difference in the relative rainfall patterns at the three sites is that there are typically 33 events per year for GED (Tables 5.1a–c), 20 events per year for KAS (Table 5.1b), and 11 events per year at KHA (Table 5.1c), implying that the differences in the annual totals at each station are not primarily due to differences in storm magnitudes (as summarized in Tables 5.1a–c and 5.5), but rather to differences in storm frequency (Tables 5.1a–c).

Generally speaking, although only daily data are available for this study of storm events in the East Sahel, the storm statistics from the region are consistent, where applicable, with those from the West Sahel, where a higher density of rain gauges has been employed, with a higher temporal (sub-hourly) resolution (D'Amato and Lebel 1998; Balme et al. 2006; Lebel and Ali 2009).

6.6 Interannual Variability in Seasonal Rainfall Patterns

The year-to-year variability of intraseasonal patterns of rainfall is manifest in the timing of the start-of-season, intraseasonal flash floods, dry spells and droughts, the timing of the end-of-season, and the duration of the interannual dry period. It stands to reason that if there are several dozen storm events per year at a site, each of which has a duration on the order of a day, and the monsoon season extends over a period of 90 to 120 days (Fig. 3.3, Tables 3.1, 3.2 and 3.3, Fig. 4.10), there is going to be a number of days during the wet season with no rainfall (e.g. Fig. 4.3). In fact, there are a number of times when the interval between storm events during the growing season exceeds 20 days (Figs. 5.11, 5.13, Table 5.1). These might reoccur on a basis of 2 years for Gedaref, 1 year for Kassala, and somewhat less often than twice a year for Khartoum, causing significant stress on local crops, which if limited to a small geographic area, are likely to fall below the resolution of merged satellite-ground-based rasterized datasets, missed by climate studies, and escape the attention of international emergency managers, such as the Famine Early Warning Systems Network (FEWS NET) operated by the United States Agency for International Development (USAID).

Whereas these intraseasonal dekad and longer (one dekad = 10 days) dry periods are often local, and escape the attention of the larger international community, the inexorable interannual hiatus of rainfall between the monsoon seasons is a well-known fact of life for the local communities. However, not-so-subtle variabilities in this phenomenon offer their own set of challenges, even to populations quite conditioned to dealing with the monsoon cycle. A significant, and generally consistent, south-to-north spatial variability is present in the interannual dry period between the last rains in the fall and the first rains in the spring (Table 5.6). This interval increases from typically 176 days at GED, to 209 days at KAS, and to 237 days at KHA. Start-of-season (SOS) is, respectively, April 30, May 7 and June 5, for GED, KAS and KHA (Table 5.6). At the end-of-season (EOS), late light rains at KAS and KHA may persist to the end of the first dekad in October, and, at GED, into the first dekad in November, with a typical variability of ±one dekad. Start of season has a variability of ±two to three dekads at all three stations, with GED having the greatest variability due to the larger number of small, early-season, pre-monsoon showers typical of more humid stations in the Sahel. At GED, 10 % of the interannual dry periods exceed 221 days; at KAS, 10 % exceed 260 days; and at KHA, 10 percent exceed 287 days (Table 5.6).

An additional consideration is the distribution of rains throughout the rainy season. Even in relatively normal years, rainfall at all three stations is intermittent and spatially incoherent. However, over a period of years, a functional representation of the expected daily rainfall can be constructed, of which the Weibull function has been particularly useful (Figs. 4.9, 4.10). Such models, being site specific, have a number of potential applications that range from climate studies to water management and rain-fed agriculture. These analytical models complement the seasonal distribution of expected rainfall based on monthly totals

(Fig. 3.3, Tables 3.1, 3.2 and 3.3). Moreover, the models for the expected daily percent of annual rainfall developed in this report (Fig. 4.10) suggest a maximum at or about the end of the first dekad in August. This is consistent with the analysis of monthly data (Fig. 3.5, Table 3.4), and the conclusion that the expected annual maximum of monthly values is distributed between August and July, with a measurable bias toward early August.

Thus, while daily data and monthly data complement each other regarding the seasonal distribution of rainfall, this report documents certain aspects of short-term variability in rainfall that can only be characterized from adequate on-going daily coverage (D'Amato and Lebel 1998). One type of variability is associated with the intraseasonal multidekadal dry periods (Fig. 5.12, Fig. 5.13); another type of variability is at the other extreme—the nature of intense, but local, storm events. Both types of variabilities are shown to occur on temporal and spatial scales for which neither the present coverage of ground-based stations nor satellite imaging is adequate (Hermance and Sulieman 2013). While international data centers are rapidly moving to address these problems, at least on broad regional scales (GCOS 2004), it is doubtful that even when present plans are fully operational, the type of data that are expected to be available will be adequate for the refined local climate assessments needed in the future. But that is a topic in itself, beyond the scope of the present report.

6.7 A View of Climate Change

6.7.1 A Long-Term Trend in Diminishing Rainfall?

Stepping back and taking a longer term view of climate change in the region, a significant theme has emerged in the literature positing a systematic long-term trend of decreasing precipitation across the Sahel. In the East Sahel, there is a general feeling among a number of environmental managers that changing climate, drought, and desertification have already accelerated the deterioration of water resources in quality and quantity (El Moghraby 2003). The collective memories of local populations and international crises managers of the deep drought of the 1970s and 1980s appear to add substance to the forecast by some of a 15 % decrease of precipitation in the coming century, resulting in a 30 % decrease in stream discharge in the major rivers of the area (UNEP 2003; Kandji et al. 2006). Some would argue that such a systematic decrease in rainfall has been in effect over the last half century, with evidence drawn from the comparison of the hydro climate of the 1950s and 1960s with recent decades (for example, Xue 1997; Elagib and Elhag 2011). If these prognostications were true, the challenges to the emerging economies of the region are formidable (Toulmin 2009).

However, in the last few years, some of the original climate forecasts by the modeling community for the next century have come under closer scrutiny, with less confidence in the predicted weather patterns for semi-arid Africa (Biasutti et al. 2008). Moreover, it is commonly acknowledged that numerical model reccnstructions of the climate of the last half century tend to substantially underestimate the interannual variability of rainfall in the Trans-African Sahel (Giannini et al. 2008). Neither of these concerns should be taken to obviate the importance of numerical model studies of African climate; rather they should underscore the important role that surface gauge data have played and should continue to play as one of the most important technical resources available for local and international planners, especially when the data are available in long, continuous time series.

For example, certain observations supporting allegations of a current and long-term "trend" of diminishing rainfall in the East Sahel are based on rain gauge data from selected stations for the last 50 years (Xue 1997; Elagib and Elhag 2011). However, using the 1950s as a tie-point to determine a present trend is not well-advised. It should be quite apparent from the annual time series for Khartoum and Kassala (Fig. 2.2) that the time period of 1950 through 1960 is simply a decade or so of higher-than-normal annual rainfall compared to the years before and after. This accords with a number of other studies of the Sahel using long time series, or at least time series that predate the 1950s wet period. These include the Sahel-wide SPI time series (Fig. 2.5) of Kraus (1977), the W. Sahel JISAO SPI (Fig. 2.6), the Trans-African composite percent departure time series (Fig. 2.10) of Nicholson (1989), and the West Sahel continental area SPI time series of Agnew (2000). Note, however, that while the annual rainfall for Khartoum and Kassala was higher during this time of the 1950s, the time series for Gedaref is much "flatter" and close to the long-term expected annual value for the station (Fig. 2.2).

6.7.2 Fitting Trendlines to Annual Data

Returning to the SPI for Gedaref from Chap. 2, Fig. 6.2 plots these data up as a simple series of points, rather than the histograms used in Chap. 2.

Figure 6.2 shows that over a 100-year period of record the "fit" of the trend-line—if accepted as valid—implies a decrease of approximately one-third of a standard deviation (sd), which at GED is approximately 140 mm. The annual rainfall is 600 mm, so the projected decrease is 42 mm over the course of the century, or about a 7 % decrease. However, the R^2 coefficient of determination suggests that the linear model only accounts for 1 % of the variance of the data, so that—as asserted throughout this report—other shorter period fluctuations dominate any suggestion of a long-term trend. This apparently holds true for possible linear trends in the SPI time series from all three of the study sites:

$$KHA(SPI) : y = -0.0059x + 11.6; R^2 = 0.034$$

$$KAS(SPI) : y = -0.013x + 25.8; R^2 = 0.17$$

$$GED(SPI) : y = -0.0034x + 6.8; R^2 = 0.011$$

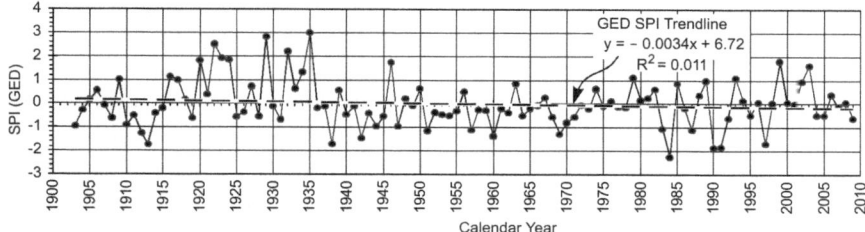

Fig. 6.2 The standardized precipitation index (SPI) for Gedaref (GED), Sudan, for the period of record available (1903–2009). Also shown is the least-squares trendline in units where x is in years and y is in standard deviations. Over a 100-year period of record, the "fit" implies a decrease of approximately one-third standard deviation. The R-squared coefficient of determination implies that more than 98 % of the interannual variance is associated with multiannual to multidecadal fluctuations

6.7.3 Comparing End-Points to Beginning-Points Over a Century of Rainfall

Another way to qualitatively identify a trend is to ask the question—for each site—is there a difference between the annual precipitation at the *beginning* of the 107-year period of record (1903–1999), that is a common base-line for the three sites, and the annual precipitation at the *end* of the period? The expected value at the beginning of each 107-year time series will be computed from the median of annual totals for the *first* decade and a half (1903–1919). Similarly the expected value at the end of the respective 107-year period of record will be computed from the median of annual totals for the *last* decade and a half (1992–2009). The respective time series used for each station will be the percent annual precipitation (Fig. 2.8).

For Khartoum (KHA), the first decade and a half of the period of record (1903–1919) has an expected (median) percent annual precipitation of $\%P(\text{KHA})|_{1903}^{1919} = 85\,\%$, which is actually less than that for the last decade and a half of the period, where $\%P(\text{KHA})|_{1903}^{1919} \approx 104\,\%$.

For Kassala (KAS), the case for a systematic difference in precipitation (Fig. 2.8) between the beginning and the end of the century of record is somewhat more compelling, with the first decade and a half (1903–1919) of the period having expected annual totals of $\%P(\text{KAS})|_{1903}^{1919} = 117\,\%$, which is somewhat greater than the last decade and a half where $\%P(\text{KAS})|_{1992}^{2009} = 86\,\%$. But most data analysts might question whether it is appropriate to fit a century-long linear trend to a dataset such as Kassala's (Fig. 2.8). The standard deviation of the 107 annual values of the $\%P(\text{KAS})|_i$ is 59 %, and the most obvious long-term pattern seems to be a half century of higher-than-expected rainfall prior to the 1960s, followed by a half century of lower-than-expected rainfall from the 1960s to present. The trendline analysis for KAS earlier in this section is consistent with a pattern of somewhat-lower rainfall at the end of the century, but more than 80 % of the interannual variance from the trend analysis is associated with shorter period interannual and decadal fluctuations.

For Gedaref (GED), evidence for a systematic difference between the beginning and the end of the epoch is not clear, with the first decade and a half (1903–1919) of the period having an expected annual percent total of $\%P(\mathrm{GED})|_{1903}^{1919} = 98\,\%$, which within the confidence of the analysis is equal to, or less than, the last decade and a half of the period where $\%P(\mathrm{GED})|_{1992}^{2009} = 105\,\%$. However the possibility of resolving a 7 % change in annual precipitation is difficult to justify, when one considers that the standard deviation of the 107 annual values of the $\%P(\mathrm{GED})|_i$ is 23 %, indicating once again the overwhelming influence of multidecadal and interannual fluctuations in modulating the hydro climate of the East Sahel.

6.7.4 The Prima Facie Evidence for Climate Change

Thus the view of climate argued for in this report—and the only one substantiated by the historical record—is that climate change is indeed occurring in the East Sahel, but at least for the hydro climate, change is not in the direction of the inexorable, monotonic downward spiral of diminishing rainfall envisaged by some. Change, in fact, is much more capricious in that, for the last century and longer, rainfall variability has lacked much trend, but has been overwhelmingly oscillatory in nature: dominated by relatively *incoherent fluctuations*, not regionally coherent *trends* of significant magnitude. What is not evident in the annual time series is any consistent, coherent pattern of monotonic change over the spatial scale of the East Sahel, and the temporal scale of the last century. For example, the present study shows that whatever long term-trend in rainfall might be present at Khartoum is sharply masked by interannual fluctuations of hundreds of percent, with a return period of less than a decade. Or, in another case, the baseline of long-term rainfall at Gedaref is virtually flat over the past 100 years, and more than 98 % of its interannual variance is associated with multiannual to multidecadal fluctuations. Perhaps most striking is the asynchronous occurrence of positive and negative extremes in annual totals among stations only a hundred or so kilometers apart, and all within the same small corner of the Sahel.

Thus, over the course of the last century, there is abundant evidence that the principal long-term characteristic of rainfall for the East Sahel is its extreme interannual and multiannual variability. The magnitude and sign of the predominant changes are not coherent among the time series from the three reference stations, and the signature of a relatively mild, century-long monotonic trend is difficult to resolve from the aliasing effects of multiannual fluctuations.

The message for those who need to mitigate the societal impact from too much or too little rainfall in the East Sahel—from water managers to agricultural producers—is one that is well known to prior generations of nomadic herders and sedentary farmers. The type of extreme temporal and spatial variability evident in the historic rainfall record unequivocally prescribes that for a culture—old or new—to be sustainable, it must be able to adapt—community-by-community—to the vagaries of climate at the local level.

6.8 Conclusions

The evidence in this report affirms that statements to the effect that there has been a coherent, regional-scale, long-term trend of diminishing rainfall in the East Sahel have little merit in the observed historical precipitation patterns. If present, such trends are poorly resolved. The type of climate change that is most evident and unequivocal in the water cycle is largely in the form of intraseasonal, interannual, decadal, and multidecadal oscillatory-type fluctuations in rainfall. The impact of this type of variability on local communities far outweighs that of any documented long-term trend of regional rainfall totals.

For purposes of merging climate science with disaster mitigation, it seems reasonable that the highest priorities should be given to assessing the greatest risks, and among these risks are (1) the year-to-year shifts in the start-of-season, (2) the timing and magnitude of the peak rainy season, (3) the duration of the interannual dry period, (4) the unpredicted intraseasonal flash floods, and (5) the dekad (10 day) and longer mid-season droughts, to name a few. The fundamental characteristic of rainfall as reported here—whether by the day, month, year, or decade—is its indeterminate spatial and temporal variability. If local communities can learn from the lessons of indigenous cultures and adapt to dealing with the range of natural variabilities reported here, they should be well-prepared to deal with the uncertainties of climate behaviors in the future. Clearly the causal relationship between the variability of local climate and regional-scale forcing in the global atmosphere needs to be better understood if climate science is to be relevant to the emerging societal needs of the East Sahel. This implies better on-the-ground observations, both in support of refined modeling and as an immediate and practical tool for local water managers.

References

Africa GIMMS NDVI (2013) Africa Global Inventory Modeling and Mapping Studies (GIMMS) 8 km 15 day composite NDVI dataset, 1981-2006. Database in Fig. 6.1 is available as AF88aug15b.n09-VIg.tif.gz at http://www.landcover.org/data/gimms/. Accessed 1 April 2013
Agnew CT (2000) Using the SPI to Identify Drought. Drought Network News (1994–2001). Paper 1 : http://digitalcommons.unl.edu/droughtnetnews/1. Accessed 29 April 2012
Ali A, Lebel T (2008) The Sahelian standardized rainfall index revisited. Int J Climatol published online in Wiley InterScience, (www.interscience.wiley.com) doi: 10.1002/joc.1832

Balme M, Vischel T, Lebel T, Peugeot C, Galle S (2006) Sahelian water balance: impact of the mesoscale rainfall variability on runoff. Part 1: rainfall variability analysis. J Hydrol 33:336–348

Bell MA, Lamb PJ (2006) Integration of weather system variability to multidecadal regional climate change: the West African Sudan-Sahel zone, 1951–98. J Clim 19(20):5343–5365

Biasutti M, Held IM, Sobel AH, Giannini A (2008) SST forcings and Sahel rainfall variability in simulations of the twentieth and twenty-first centuries. J Clim 21:3471–3486

D'Amato N, Lebel T (1998) On the characteristics of the rainfall events in the Sahel with a view to the analysis of climatic variability. Int J Climatol 18:955–974

El Gamri T, Saeed AB, Abdalla AK (2009) Rainfall of the Sudan: Characteristics and Prediction. Arts J 27: 18–35. J Faculty Arts, Univ of Khartoum, Sudan. http://adabjournal.uofk.edu/current%20issue/ISSUES%20ENGLISH/El%20Gamri_%20Amir%2. Accessed 24 February 2013

El Moghraby AI (2003) State of the environment in Sudan. UNEP Studies of EIA Practice in Developing Countries, Case Study 4: 27- 36. ISBN: 92 807 2298 0. http://www.unep.ch/etu/publications/Compendium_toc.htm. Accessed 15 Mar 2013

El Tom MA (1975) The rains of the Sudan: mechanisms and distribution. Khartoum University Press, Khartoum

Elagib NA, Elhag MM (2011) Major climate indicators of ongoing drought in Sudan. J Clim 409(3–4):612–625. doi:10.1016/j.jhydrol.2011.08.047

GCOS (2004) Implementation Plan for the Global Observing System for Climate in Support of the UNFCCC, Executive Summary, October 2004, GCOS – 92 (ES), (WMO/TD No. 1244). Http://www.wmo.int/pages/prog/gcos/Publications/gcos-92_GIP_ES.pdf. Accessed 10 October 2012

Giannini A, Saravanan R, Chang P (2003) Oceanic forcing of Sahel rainfall on interannual to interdecadal time scales. Science 302:1027–1030

Giannini A, Biasutti M, Verstraete MM (2008) A climate model-based review of drought in the Sahel: desertification, the re-greening and climate change, Global and Planetary Change 64(3–4): 119–128, ISSN 0921-8181. doi: 10.1016/j.gloplacha.2008.05.004

Held IM, Delworth TL, Lu J, Findell KL, Knutson TR (2005) Simulation of Sahel drought in the 20th and 21st centuries. Proceedings of the United States National Academy of Science 102(50):17891–17896

Hermance JF, Sulieman HM (2013) Comparing satellite RFE data with surface gauges for 2012 extreme storms in African East Sahel. Remote Sen Lett 4(7):696–705. doi:10.1080/2150704X.2013.787498

Hulme M (2001) Climate perspectives on Sahelian desiccation: 1973–1998. Global Environ Change 11(1):19–29. doi:10.1016/S0959-3780(00)00042-X

Jurkovica RS, Pasaric Z (2012) Spatial variability of annual precipitation using globally gridded data sets from 1951 to 2000. Int J Climatol. doi:10.1002/joc.3462

Kandji T, Verchot L, Mackensen J (2006) Climate Change and Variability in the Sahel Region: Impacts and Adaptation Strategies in the Agricultural Sector. United Nations Environment Programme (UNEP), Word Agroforestry Centre (ICRAF), Nairobi, Kenya. http://worldagroforestrycentre.net/. Accessed: 4 July 2012

Kraus EB (1977) Subtropical droughts and cross-equatorial energy transports. Mon Weather Rev 105:1009–1018

Lebel T, Ali A (2009) Recent trends in the Central and Western Sahel rainfall regime (1990–2007). J Hydro 375:52–64

Nicholson SE (1986) The spatial coherence of African rainfall anomalies: Interhemispheric teleconnections. J Clim Appl Meteorol 25:1365–1381

Nicholson SE (1989) Long term changes in African rainfall. Weather 44:46–56. doi:10.1002/j.1477-8696.1989.tb06977

Nicholson SE (2011) Dryland climatology. Cambridge University Press, Cambridge

Nicholson SE, Selato JC (2000) The influence of La Nina on African rainfall. Int J Climatol 20(14):1761–1776

Toulmin C (2009) Climate change in Africa. Zed Books, London

Tucker CJ, Pinzon JE, Brown ME, Slayback D, Pak EW, Mahoney R, Vermote E, Saleous N (2005) An extended AVHRR 8-km NDVI data set compatible with MODIS and SPOT vegetation NDVI data. Int J Remote Sens 26(20):4485–4498

UNEP (2003) Sudan: Post-Conflict Environmental Assessment. Chap. 3: Natural disasters and desertification. UNEP Disasters and Conflicts Programme. http://postconflict.unep.ch/publications/sudan/ Accessed 17 February 2013

Xue Y (1997) Biosphere feedback on regional climate in tropical North Africa. Quart J Roy Met Soc 123:1483–1515

Summary

The northward migration of the African monsoon rains in summer, associated with the seasonal march of the Intertropical Convergence Zone (ITCZ) across the plains south of the Sahara, is the most critical asset for the livelihoods of indigenous peoples and local economies of the Sahel. It is essential that climate science (and its publicly available database) play a key role in characterizing the variabilities of these rainfall patterns in space and time if sustainable life styles are to accommodate the expanding populations of the region. This study turns to the East Sahel of Sudan by analyzing over 100 years of historical rainfall data from three of the few long term standard WMO rain gauge stations in substantially different rainfall settings. From north to south, transecting the Sahel, the stations with their annual rainfall are Khartoum (130 mm); Kassala (280 mm); and Gedaref (600 mm). The conclusions challenge a popular notion that changing climate, drought and desertification in the East Sahel may have already accelerated the deterioration of its water resources. However, any evidence of a persistent and coherent regional trend of diminishing rainfall is obscure. Quite the contrary, the evidence demonstrates that the fluctuations of climate and weather patterns over the ensuing decades of the past century—at all temporal scales from days to years to decades—profoundly overwhelm any suggestion of a large-scale, coherent decrease (or increase) in rainfall. The implication is that, it is not long term change, but the highly localized interseasonal, interannual and multiannual variability of rainfall that poses the greatest and most immediate societal threat from naturally-induced causes; a process constantly destabilizing an agrarian economy struggling to survive in a climate that irregularly vacillates between years of drought and years of flooding. While this report may have some interest for climate scientists, it is primarily directed to a general readership (including students in public policy and anthropology) concerned with the availability of water in the Sahel, particularly the long term sustainability of local small-scale farms and transhumant pastoralism.

John F. Hermance
Brown University
e-mail: John_Hermance@Brown.Edu

J. F. Hermance, *Historical Variability of Rainfall in the African East Sahel of Sudan*, SpringerBriefs in Earth Sciences, DOI: 10.1007/978-3-319-00575-1, © The Author(s) 2014